New Trends in Telecommunications Engineering

New Trends in Telecommunications Engineering

Editor

Jaume Anguera

MDPI

Basel • Beijing • Wuhan • Barcelona • Belgrade • Novi Sad • Cluj • Manchester

Editor
Jaume Anguera
Universitat RamonLlull
Barcelona
Spain

Editorial Office
MDPI
St. Alban-Anlage 66
4052 Basel, Switzerland

This is a reprint of articles from the Special Issue published online in the open access journal *Applied Sciences* (ISSN 2076-3417) (available at: https://www.mdpi.com/journal/applsci/special_issues/telecommunications_engineering).

For citation purposes, cite each article independently as indicated on the article page online and as indicated below:

Lastname, A.A.; Lastname, B.B. Article Title. *Journal Name* **Year**, *Volume Number*, Page Range.

ISBN 978-3-7258-0993-6 (Hbk)
ISBN 978-3-7258-0994-3 (PDF)
doi.org/10.3390/books978-3-7258-0994-3

Contents

About the Editor

Jaume Anguera

Dr. Jaume Anguera, IEEE Fellow, founder and CTO at Ignion. Associate Prof. at Ramon LLull University. Inventor of more than 180 granted patents, licensed to telecom companies. Inventor of Antenna Booster Technology, which fostered the creation of Ignion. Author of + 300 scientific papers and international conferences (h-index 56), and 8 books. He has participated in + 24 competitive public funded research projects exceeding EUR 14M,. He has taught + 80 antenna courses worldwide. With 25 years of R&D experience, he has developed part of his professional experience with Fractus in South Korea in the design of miniature antennas for large Korean companies. He has received several national and international awards. He has directed the master/doctorate thesis of +180 students. His biography appears in Who'sWho in the World and Who'sWho in Science and Engineering. He is associate editor of the IEEE Open Journal Antennas Prop., Elec. Letters. He is an IEEE Antennas Prop. and a distinguished Lecturer.

Preface

This collection of research papers explores various aspects of antenna design and high-speed wireless communication, delving into the forefront of technology for modern telecommunications. Antennas are crucial components in wireless systems, enabling the transmission and reception of data. The papers in this collection investigate innovative antenna designs that cater to the growing demands of future-generation wireless applications, including 5G networks, Internet of Things (IoT) devices, and ultra-wideband (UWB) communication.

These papers discuss antennas with characteristics such as super-wideband capabilities, dual-band operation, high gain, and multiple-input multiple-output (MIMO) configurations. Additionally, they explore novel techniques in order to minimize interference and optimize antenna layouts for improved performance. Furthermore, this collection delves into topics related to wireless communication efficiency, including error reduction methods and advanced optical nano-antenna designs for photonics applications. Overall, these papers represent the ongoing efforts to push the boundaries of antenna technology and enhance high-speed wireless communication systems for a connected and data-driven future.

Jaume Anguera
Editor

applied sciences

MDPI

Editorial

Special Issue on New Trends in Telecommunications Engineering

Jaume Anguera [1,2]

[1] R&D Department at Ignion, 08190 Barcelona, Spain; jaume.anguera@salle.url.edu
[2] Smart Society Research Group, Universitat Ramon LLull, 08022 Barcelona, Spain

This collection of research papers explores various aspects of antenna design and high-speed wireless communication, delving into the forefront of technology for modern telecommunications. Antennas are crucial components in wireless systems, enabling the transmission and reception of data. The papers in this collection investigate innovative antenna designs that cater to the growing demands of future-generation wireless applications, including 5G networks, Internet of Things (IoT) devices, and ultra-wideband (UWB) communication.

These papers discuss antennas with characteristics such as super-wideband capabilities, dual-band operation, high gain, and multiple-input multiple-output (MIMO) configurations. Additionally, they explore novel techniques in order to minimize interference and optimize antenna layouts for improved performance. Furthermore, this collection delves into topics related to wireless communication efficiency, including error reduction methods and advanced optical nano-antenna designs for photonics applications. Overall, these papers represent the ongoing efforts to push the boundaries of antenna technology and enhance high-speed wireless communication systems for a connected and data-driven future.

A brief summary of the papers can be found below.

Ref. [1] paper proposes a high-dimension ratio, octagonal-shaped, super-wideband (SWB) monopole antenna. The antenna exhibits a frequency range of 3.71 to 337.88 GHz with good performance metrics such as $|S11| \leq -10$ dB, VSWR < 2, a bandwidth ratio (BR) of 91.07:1, and a very high bandwidth dimension ratio (BDR) of 6057.27. The antenna is designed on a Rogers 5880 dielectric substrate with compact dimensions of $14 \times 16 \times 0.787$ mm^3. The simulated and measured results show good agreement, making the antenna suitable for future-generation mobile networks due to its strong radiation properties, compactness, and extremely wide bandwidth.

Ref. [2] presents a design of an elliptical-shaped dual-band UWB notch antenna for wireless applications. The antenna incorporates corner cuts on a partial ground plane to achieve a UWB bandwidth and utilizes inverted-U-shaped and conductor-shaped resonators for dual-band notch characteristics. The proposed antenna has dimensions of 24×32 mm^2 and offers a gain of 4.9 dB, a bandwidth of 2.5–11 GHz, and a stable radiation pattern. It rejects WLAN and ITU bands from 5.2 to 5.7 GHz and 7.2 to 8.5 GHz, respectively. The article discusses the evolution stages of the UWB antenna and the parametric analysis of the UWB notch antenna.

Ref. [3] presents a compact antenna system with a multiple-input multiple-output (MIMO) configuration for mm-wave 5G-based Internet of Things (IoT) applications. The antenna elements are arranged to achieve significant field decorrelation and a dielectric layer is added to improve the radiation characteristics. The MIMO configuration has a size of 14 mm $\times$ 20 mm and operates in the frequency range of 16.7 to 25.4 GHz. It achieves a peak gain of 8.47 dB, more than 80% total efficiency, and high isolation of over -30 dB. The design demonstrates wideband features, improved gain, and high isolation, making it suitable for mm-wave-based 5G applications.

Citation: Anguera, J. Special Issue on New Trends in Telecommunications Engineering. *Appl. Sci.* **2023**, *13*, 9892. https://doi.org/10.3390/app13179892

Received: 21 August 2023
Accepted: 29 August 2023
Published: 31 August 2023

Ref. [4] explains a novel MIMO antenna system for ultra-wideband (UWB) applications is presented. The system consists of a 4×4 array of antenna elements arranged in a plus-shaped configuration. The individual antenna element is a fractal circular ring structure with a modified partial ground plane. It operates in the frequency range of 2.67 GHz to 12 GHz, with a wide impedance bandwidth of 9.33 GHz. The system provides pattern and polarization diversity without the need for any isolation enhancement network. The fabricated and tested MIMO antenna shows UWB response and good isolation between the antenna elements, making it suitable for UWB MIMO applications.

Ref. [5] proposes efficient helical and FRF interleavers for OFDM systems, enhancing reliability and throughput. The study compares FFT-OFDM with DWT-OFDM using various wavelet families and channel models. The results show DWT-OFDM outperforms FFT-OFDM, particularly with the Haar wavelet. The simulation indicates FRF interleaving improves the BER for FFT-OFDM, while DWT-OFDM's FRF interleaver performs similarly to the helical interleaver. The recommendation is to use FRF interleaving for FFT-OFDM, improving energy efficiency. Further research is needed to compare conventional interleavers with the proposed FRF interleaver.

Ref. [6] presents a broadband optical nano-antenna with a hybrid plasmonic feed, designed for nano-photonic applications. The antenna exhibits a gain of up to 11.4 dBi and covers a wide range of optical communication wavelengths. It features a unique hybrid plasmonic waveguide-based feed that enables directional radiation properties. The antenna is suitable for inter- and intra-chip optical communications and sensing applications. The study also explores array configurations to increase gain and directionality, making it useful for optical energy harvesting.

Ref. [7] presents the optimization of a coherent dual-beam array feed network for aperiodic concentric ring antennas. The feeding system is based on a reconfigurable topology with alternated power dividers and combiners, providing coherent in-phase outputs. The study analyzes a two-beam architecture with multi-beam shaping and steering features, and optimizes the aperiodic layout based on the radii of the circular rings using the differential evolution method. The numerical experiments validate the improved performance of the proposed dual-beam configuration with a non-uniform layout compared to its uniform counterpart. The results show an enhanced performance in terms of sidelobe level and directivity.

Ref. [8] studies a compact concentric structured wideband antenna with triple notch bands was designed and experimentally demonstrated. The antenna exhibited a broad impedance bandwidth of 45.83 GHz (1.67–47.5 GHz) with three notches at 1.8–2.2 GHz, 4–7.2 GHz, and 9.8–10.4 GHz to minimize interference from the AWS, C, and X bands, respectively. The simulated and measured results of the VSWR, gain, and radiation pattern showed good agreement. The antenna also demonstrated excellent time-domain characteristics, including minimal pulse distortion and constant group delay. A comparative analysis with existing designs showed that the proposed antenna outperformed in terms of bandwidth, gain, and rejection levels. Overall, the designed antenna is a promising solution for pulse-based wideband communication systems.

Ref. [9] describes a modified Franklin array antenna for 5G wireless applications operating at 22.7 and 34.9 GHz. The antenna consists of a 3×3 array patch element and a slotted ground plane, designed for broad bandwidth, high directivity, and dual-band operation. The dimensions of the patch antennas are based on $\lambda/2$ of the second resonant frequency. The simulation and measurement results show a dual-band operation with good impedance bandwidth. The antenna performance is analyzed through parametric analysis, and the design is carried out using HFSS v.14.0. The proposed antenna overcomes the limitations of previous collinear arrays and Franklin antennas, offering wideband operation, a compact size, and improved performance.

Conflicts of Interest: The author declares no conflict of interest.

References

1. Suguna, N.; Revathi, S. A Compact Super-Wideband High Bandwidth Dimension Ratio Octagon-Structured Monopole Antenna for Future-Generation Wireless Applications. *Appl. Sci.* **2023**, *13*, 5057. [CrossRef]
2. Kumar, O.P.; Ali, T.; Kumar, P.; Kumar, P.; Anguera, J. An Elliptical-Shaped Dual-Band UWB Notch Antenna for Wireless Applications. *Appl. Sci.* **2023**, *13*, 1310. [CrossRef]
3. Sehrai, D.A.; Asif, M.; Khan, J.; Abdullah, M.; Shah, W.A.; Alotaibi, S.; Ullah, N. A High-Gain and Wideband MIMO Antenna for 5G mm-Wave-Based IoT Communication Networks. *Appl. Sci.* **2022**, *12*, 9530. [CrossRef]
4. Alharbi, A.G.; Rafique, U.; Ullah, S.; Khan, S.; Abbas, S.M.; Ali, E.M.; Alibakhshikenari, M.; Dalarsson, M. Novel MIMO Antenna System for Ultra Wideband Applications. *Appl. Sci.* **2022**, *12*, 3684. [CrossRef]
5. El-Fouly, F.H.; Ramadan, R.A.; Abd El-Samie, F.E.; Kachout, M.; Alzahrani, A.J.; Alshudukhi, J.S. Burst Channel Error Reduction Based on Interleaving for Efficient High-Speed Wireless Communication. *Appl. Sci.* **2022**, *12*, 3500. [CrossRef]
6. Ahmad, I.; Ullah, S.; Din, J.U.; Ullah, S.; Ullah, W.; Habib, U.; Khan, S.; Anguera, J. Maple-Leaf Shaped Broadband Optical Nano-Antenna with Hybrid Plasmonic Feed for Nano-Photonic Applications. *Appl. Sci.* **2021**, *11*, 8893. [CrossRef]
7. Arce, A.; Stevens-Navarro, E.; Pineda-Rico, U.; Cardenas-Juarez, M.; Castillo-Soria, F.R.; Covarrubias, D.H. Optimization of a Coherent Dual-Beam Array Feed Network for Aperiodic Concentric Ring Antennas. *Appl. Sci.* **2021**, *11*, 1111. [CrossRef]
8. Balani, W.; Sarvagya, M.; Ali, T.; Samasgikar, A.; Das, S.; Kumar, P.; Anguera, J. Design of SWB Antenna with Triple Band Notch Characteristics for Multipurpose Wireless Applications. *Appl. Sci.* **2021**, *11*, 711. [CrossRef]
9. Surendran, A.; Ali, T.; Kumar, O.P.; Kumar, P.; Anguera, J. A Dual-Band Modified Franklin mm-Wave Antenna for 5G Wireless Applications. *Appl. Sci.* **2021**, *11*, 693. [CrossRef]

applied sciences

MDPI

Article

A Dual-Band Modified Franklin mm-Wave Antenna for 5G Wireless Applications

Arjun Surendran [1], Aravind B [1], Tanweer Ali [1,*], Om Prakash Kumar [1], Pradeep Kumar [2,*] and Jaume Anguera [3,4]

1 Department of Electronics and Communication, Manipal Institute of Technology, Manipal Academy of Higher Education, Manipal 576104, India; arjun.surendran@learner.manipal.edu (A.S.); aravind.b@learner.manipal.edu (A.B.); omprakash.kumar@manipal.edu (O.P.K.)
2 Discipline of Electrical, Electronic and Computer Engineering, University of KwaZulu-Natal, Durban 4041, South Africa
3 Fractus Antennas, 08174 Barcelona, Spain; jaume.anguera@salle.url.edu
4 Electronics and Telecommunication Department, Universitat Ramon LLull, 08022 Barcelona, Spain
* Correspondence: tanweer.ali@manipal.edu (T.A.); kumarp@ukzn.ac.za (P.K.)

Abstract: Franklin array antennas are considered as one of the most competitive candidates for millimeter-wave (mmW) 5G applications due to their compact size, simple geometry and high gain. This paper describes a microstrip Franklin antenna array for fifth generation (5G) wireless applications. The proposed modified Franklin array is based on a collinear array structure with the objective of achieving broad bandwidth, high directivity, and dual-band operation at 22.7 and 34.9 GHz. The designed antenna consists of a 3 × 3 array patch element as the radiating part and a 3 × 3 slotted ground plane operating at a multiband resonance in the mmW range. The dimensions of the patch antennas are designed based on $\lambda/2$ of the second resonant frequency. The designed antenna shows dual band operation with a total impedance bandwidth ranging from 21.5 to 24.3 GHz (fractional bandwidth of 12.2%) at the first band and from 33.9 to 36 GHz (fractional bandwidth of 6%) at the second band in simulation. In measurement, the impedance bandwidth ranges from 21.5 to 24.5 GHz (fractional bandwidth of 13%) at the first band and from 34.3 to 36.2 GHz (fractional bandwidth of 5.3%) at the second band, respectively. The performance of the antenna is analyzed by parametric analysis by modifying various parameters of the antenna. All the necessary simulations are carried out using HFSS v.14.0.

Keywords: collinear array; dual band; Franklin antenna; millimeter-wave (mmW)

Citation: Surendran, A.; B, A.; Ali, T.; Kumar, O.P.; Kumar, P.; Anguera, J. A Dual-Band Modified Franklin mm-Wave Antenna for 5G Wireless Applications. *Appl. Sci.* **2021**, *11*, 693. https://doi.org/10.3390/app11020693

Received: 13 December 2020
Accepted: 5 January 2021
Published: 12 January 2021

Publisher's Note: MDPI stays neutral with regard to jurisdictional claims in published maps and institutional affiliations.

1. Introduction

The evolution of mobile standards from the first generation (1G) standard to the fifth generation (5G) standard, where we currently stand, and the future standards is considered as a revolution in the field of wireless communications. It all began with 1G, where the basic calling facility was introduced with no proper coverage and security for the network. 2G networks changed the whole concept with the introduction of a messaging facility or short message system as well as with digital voice calls and encryption of the calls. The second-generation network laid the foundation for all wireless communication networks. The introduction of 3G, or the packet-switched networks improved data rates, increased data transfer capability, as well as the introduction of video calls for applications such as video streaming and video conferencing revolutionized the mobile standards [1]. Fourth-generation (4G) networks are currently the most popular network standard in the present world. The transition from the previous standard to the present standard required various infrastructure changes to the antenna as well as the other equipment such that it can become 4G compatible. All the above standards face a common problem, which is the scarcity of bandwidth. The 5G standard or the internet of things era is the solution that overcomes this problem. 5G technology requires new infrastructure that has to be laid for the implementation. It is a mature technology that has proven to be of high data rate communication [2,3].

The wireless spectrum that exists presently is losing its free space, which has led to researchers exploring the unused millimeter-wave (mmW) spectrum for such applications [4]. Millimeter wave technology exploits the frequency spectrum from 30 to 300 GHz. The wavelength required for the operation of mmW varies from 10 to 1 mm. The advantages of using the above technology include high data rates, almost the same as those obtained in optical fibers, achieved at a lesser cost or free of cost [5,6]. The added merits to this technology are that at high frequencies, the traffic is less, and the communication would be more secure because these frequencies cannot be utilized by everyone. The drawbacks are that a completely modified infrastructure has to be laid out for obtaining these higher frequencies [7,8]. Proper substrate selection is needed with an appropriate value of dielectric constant, while the thickness of the substrate also plays an important role in the deployment of this technology.

mmW technology deals with frequencies in the range of GHz, so the conventional single-antenna structures might fall short of the requirements such as high gain, directive patterns, and bandwidth. So, the need of combining of antenna structures to form array structures is required. Among the different types of antenna array structures, one of the array types is the collinear array (CoA) antenna. A series of antenna elements that are fixed in such a manner that they are parallel and located along a common axis is termed a collinear antenna array. This antenna principle is further modified and applied to a new antenna termed Franklin antenna [9]. In a Franklin array antenna, in addition to the antenna patch elements that are placed in a collinear manner, two transmission lines are placed in between the patches as well as a phasing stub assembly. High gain is a feature of these antennas. The restriction of these classes of antennas is that they are mainly restricted to a single band of operation [9–11].

Several researchers showed interest in the collinear array principle of antennas in improving the various antenna parameters. A simple arrangement of collinear microstrip patch antenna is illustrated in [12,13], which was limited to a single band operation with a high gain. The narrowband, along with single-band operation, was the limitation of this antenna. Various modifications of the collinear array antenna were introduced, of which the Franklin antenna is an example. Researchers have worked on several modifications of this antenna to obtain improved antenna parameters. One such work is demonstrated in [14], where it is used for radar application. Here, the conventional Franklin antenna is modified for automotive short-range radar applications in which the phase-shifting sections are arranged in a simple straight-line configuration. It is again restricted to a single band operation with less bandwidth. Another variation of the Franklin antenna is presented in [15], which is used for wireless charging applications. The design is complex, as it consists of 32 elements arranged in a collinear array pattern. Since it is designed for wireless charging applications, the frequency of operation is low. A 14 element Franklin antenna was proposed in [16]. The advantage of this design was that a simplified feeding technique was introduced, which gave rise to a more longitudinally and transversely compact class of antennas. A circularly polarized Franklin antenna is demonstrated in [17], where it finds its application in radars. Here the 12 elements are arranged circularly to obtain a circular class of Franklin antenna. The design provides high inter port isolation and moderate impedance. All these antennas were limited to narrowband operation, and in [18], a modified Franklin antenna was introduced where the upper and lower cut off frequencies are designed based on the patch length and stub length. The six-element array antenna generated a wide bandwidth over a single resonant frequency. In [19], the Franklin antenna was developed on dielectric slab to operate at single band of operation. Here, the meander-based geometry is used to obtain high impedance phenomenon at the desired single band resonance. The designed antenna is powered by standard and surface waves along with a dielectric structure, in such a way that the energy is gradually radiated. Finally, in [20], the design that was used for the multi-band response was modified to obtain a single band operation. The design was modified in such a way that additional uniform slots were introduced in the ground plane. This enabled the antenna to resonate at a single frequency with a suitable bandwidth. Even though this antenna results in a large bandwidth as well as gain, the drawback was that it was limited to a single frequency of operation.

All of the collinear array antennas and Franklin antennas that were designed earlier faced certain drawbacks, such as the single/dual resonant frequency with narrowband operation, limited bandwidth, low directive pattern, and complex structure. Due to these issues, it may be difficult for them to be integrated with portable wireless devices for modern wireless communication systems. In this paper, the proposed antenna introduces the concept of a Franklin antenna that has been modified in such a way to obtain a wideband antenna resonating at two different frequencies i.e., 22.7 and 34.9 GHz. The proposed antenna is designed in four steps. In the first step, the Franklin radiator is designed by utilizing a 3 × 3 unit cell array structure with full ground plane. This structure resulted in four mmW bands with poor impedance matching at the first three bands. So, in order to overcome this problem, the ground plane is etched with a one-by-three array slot format in the second step. This modification also resulted in triple band operation but with poor impedance matching at the first two bands. In the third step, the slot configuration at the ground part is modified in the form of a two-by-three array, while keeping the radiator unchanged. This configuration resulted in dual band operation with S_{11} very close to −10 dB, at both bands. Finally, in the fourth step, in order to get the desired wide bandwidth and good impedance matching with dual band operation (i.e., at 22.7 and 34.9 GHz), the ground part is further etched with a slot configuration of a three-by-three array. A detailed parametric investigation is also carried out to fix the dimension of the proposed design. The main advantage of the antenna is that the aforementioned modifications are introduced into the radiating part and ground plane in such a manner that there is no additional increase in the area of the antenna, i.e., the antenna remains compact.

2. Antenna Design Approach

2.1. Proposed Antenna Configuartion and Dimension Layout

The final antenna or the proposed antenna with detailed optimized dimensions is shown in Figure 1. The antenna has a 3 × 3 array-like structure, which consists of 9 patch elements, of which six of these patch elements have a dimension of $P_L \times P_W$. Three of them have a dimension of $(P_L + R) \times P_W$. Each row of these patch elements also has folded dipole elements that connect the row-wise patch elements. The antenna feeding network is provided to the last row of the patch elements. The ground plane consists of nine slots that are etched in a 3 × 3 manner to provide the required response. The last cut in the first column of the array follows a slight misalignment to obtain the proper response with an appropriate gain required for proper operation. The detailed dimensions of the antenna are shown in Table 1. The antenna designed in this paper is based on the concept of [7,11,18,20].

The proposed modified Franklin array structure is based on a collinear array structure. A set of patches and stubs constitutes a collinear array structure. The modified Franklin array structure consists of 9 patch antennas arranged in a 3 × 3 configuration, and in between each row of these patches contains stubs that have been changed to symmetrically folded dipole like structures, as illustrated in Figure 1. These 9 patch antennas and folded dipole-like structures arranged in a 3 × 3 array manner are connected to a feeding network that has been modified into a folded dipole like structure and connected from the last row of patch antennas and folded dipole like structures.

The patch antennas are of dimensions $P_L \times P_W$. The six patch antennas in the second and third row of the array follow these dimensions. The first row of patch antennas has an incremental length R added to it. The dimensions of the patch antennas are designed based on $\lambda/2$ of the second resonant frequency. Similarly, in an arrangement similar to the 3 × 3 arrangement of patch elements, modifications are made to the ground in a 3 × 3 manner to the original design such that the desirable dual-band operation of the antenna is achieved.

HFSS, or High-Frequency Structure Simulator software, is used in the modeling, simulation, and analysis, as well as for the numerical evaluation of the antenna design. The substrate used had a height (h) of 0.8 mm, and a metallic ground was used at the bottom.

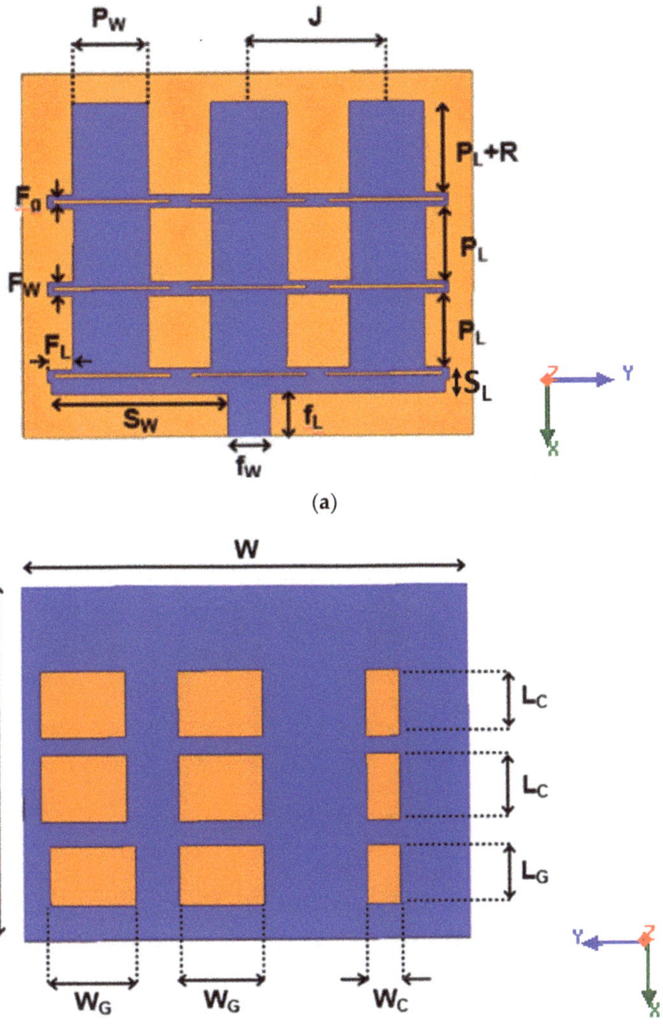

Figure 1. Final antenna (**a**) top view and (**b**) bottom view.

Table 1. Optimized dimensions of the final proposed antenna.

Symbol	Parameters	(mm)
P_L	Patch element length	4.34
P_W	Patch element width	4.3
R	Patch element length extension for the first row	1.0
J	Spacing between the centers of patch elements in a row	8.01
F_L	Folded dipole length	1.4

Table 1. *Cont.*

Symbol	Parameters	(mm)
F_W	Folded dipole width	0.7
S_W	Matching stub width	10.15
S_L	Matching stub length	1.4
f_L	Feed length	3.2
f_W	Feed width	2.4
F_g	The gap of the folded dipole	0.26
L	Length of substrate material	21.0
W	Width of substrate material	26.0
W_G	Width of first type ground cut	5.0
W_C	Width of second type ground cut	2.0
L_G	Length of first type ground cut	3.6
L_C	Length of second type ground cut	4.0

2.2. Design Steps to Obtain the Proposed Antenna

In this section, the procedure to obtain the final antenna is described in detail. The proposed Franklin antenna array is designed in four steps. Initially, the first antenna is designed with a full ground plane at the bottom of the substrate and the top portion of the substrate consisting of the patch consisting of a 3 × 3 array of a unit cell. This antenna is termed as antenna M. The representation of the antenna structure is shown in Figure 2a,b. This antenna shows a quad band of operation at 24.4, 25.8, 26.6 and 32.7 GHz with very poor impedance matching at the first three bands.

Figure 2. (**a**). Top view of the antenna (M) showing the patch array structure of the antenna, (**b**) bottom view of the antenna showing the full ground plane.

So, to improve the impedance matching in the initial antenna M, in the next step, slots are introduced in the ground plane. This antenna is now termed antenna N. The representation of the antenna structure is shown in Figure 3a,b, wherein it can be seen that three cuts are made in the ground plane. Two cuts are of dimensions $W_G \times L_G$, and one cut is of dimensions $W_C \times L_G$. The last column's squares are made with different sizes and are not of the same dimensions as the other squares, in order to make the antenna resonate at a particular frequency. This modification also leads to a triple band of operation at 22.5, 24.9 and 35.3 GHz, with very poor impedance matching at the last two bands.

To further improve the performance of the designed antenna N, in the next stage an additional three slots are etched to the ground part. This modified antenna is now termed antenna O. The representation of this antenna structure is shown in Figure 4a,b. The slots are now of a 2 × 3 array form. The new row of slots formed consists of two cuts of dimensions $W_G \times L_C$ and one cut of dimensions $W_C \times L_C$. The last cut that lies close to the boundary of the ground is slightly misaligned to the cut made in the previous stage for obtaining a proper response. This modification allows the antenna to operate at dual-band i.e., at 22.6 and 35.05 GHz, with a lower reflection coefficient value at both the bands.

(**a**) (**b**)

Figure 3. (**a**). Top view of the antenna (N) showing the patch array structure of the antenna, (**b**) bottom view of the antenna showing the ground plane with the first set of three cuts.

(**a**) (**b**)

Figure 4. (**a**) Top view of the antenna (O) showing the patch array structure of the antenna, (**b**) bottom view of the antenna showing the ground plane with the second set of three cuts.

To further improve the impedance matching at the dual band, the final modifications are made to the ground part of antenna O by etching one more row of square slots, thereby resulting in 9 cuts in the ground plane. The introduction of these slots further affect the surface current behavior of the antenna, thereby improving the impedance matching of the antenna. This antenna is now termed as P (proposed antenna). The representation of the antenna structure is shown in Figure 5a,b. The cuts that were introduced in the ground plane are in the form of a 3 × 3 array structure. The last row of cuts formed consists of two cuts of dimensions $W_G \times L_C$ and one cut of dimensions $W_C \times L_C$. The last cut that lies close to the ground's boundary is slightly misaligned to the cut made in the first stage and aligned to the cut made in the previous stage for obtaining a proper frequency response for the required dual-band operation. Thus, one of the cuts in the final proposed antenna is misaligned to the other cuts of the same column. The reflection coefficient of the antenna's various stages is illustrated in Figure 6. It can be observed that the antenna offers a dual-band operation at 22.7 and 34.9 GHz with excellent impedance matching at both the bands. Thus, it can be concluded that the slots are added to the ground plane for the antenna to radiate at the particular frequencies and improve the input impedance matching and S_{11}.

Figure 5. (**a**). The top view of the antenna showing the antenna's patch array structure, (**b**) the bottom view of the antenna showing the ground plane with the third set of three cuts.

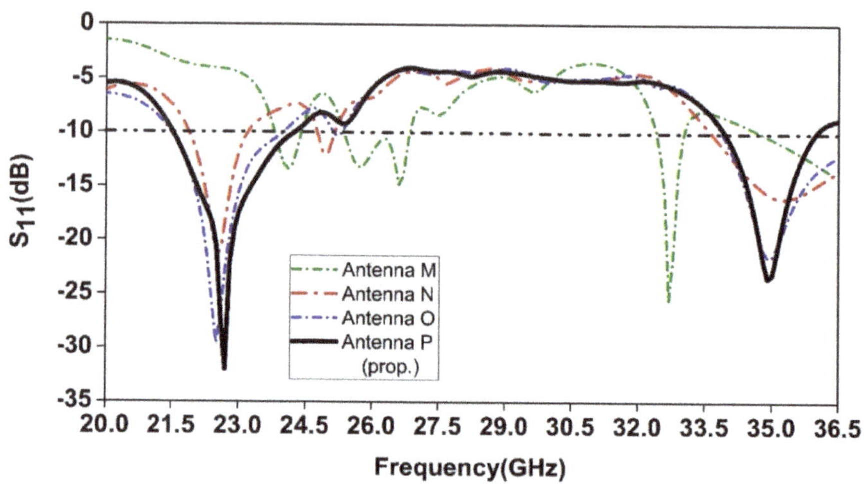

Figure 6. Variation in reflection coefficient with frequency for different antenna configurations.

3. Parametric Analysis

A parametric study analyzes the patch element's impact, and the slots etched in the ground plane on the antenna. The design's performance is mainly influenced by the incremental length (R), the patch element length (P_L), the length and width of the slots in the ground plane L_G, L_C, W_G, W_C.

3.1. Effect of Length of Patch Element P_L

The effect of the length of the patch element on the antenna is studied. The length of the element is increased to analyze its effect. The increase in the length causes a variation in the resonant frequency as well as the return loss of the antenna. The variation of the length of patch P_L is shown in Figure 7.

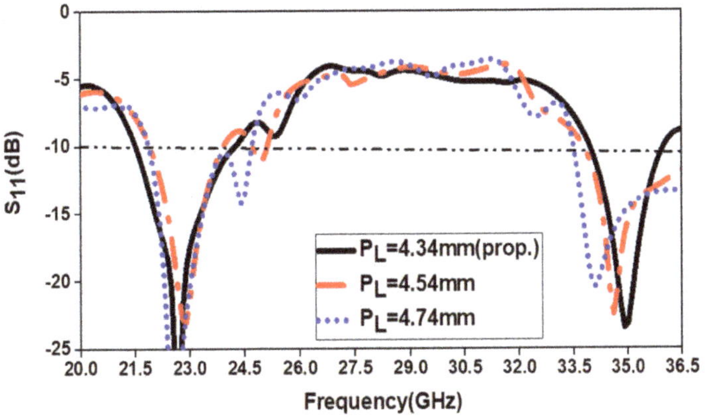

Figure 7. Analysis of variation in the length of patch element P_L.

3.2. *Effect of Incremental Length R*

The effect of the incremental length of the patch elements of the first row on the antenna is analyzed. The length is varied to analyze its effect. The variation in the length does not cause a significant variation in the resonant frequency, but a very significant variation in the antenna's return loss is seen. The variation of the incremental length R is demonstrated in Figure 8.

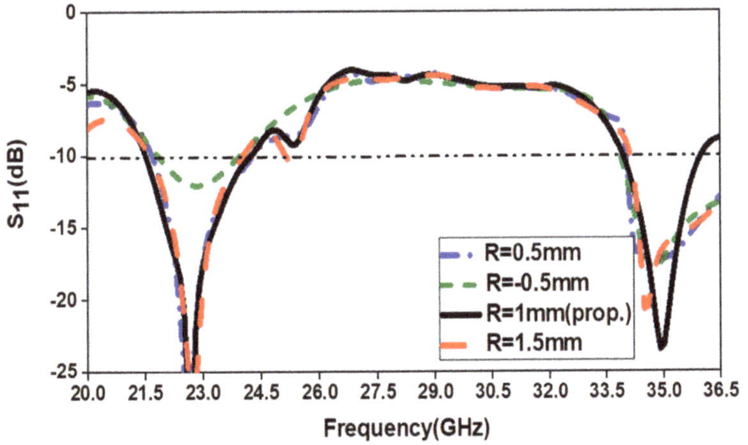

Figure 8. Analysis of variation in the incremental length R.

3.3. *Effect of the Width of the First Type of Ground Cut W_G*

The effect of the width of the first type of ground cut on the antenna is studied. The width is varied to study its impact. The variation in the width causes a significant variation in both the resonant frequency, but a very substantial variation in the antenna's return loss is observed at the first resonant frequency. At the second variation, not much change is observed in terms of return loss. The variation of the width of the first type of ground cut W_G is given in Figure 9.

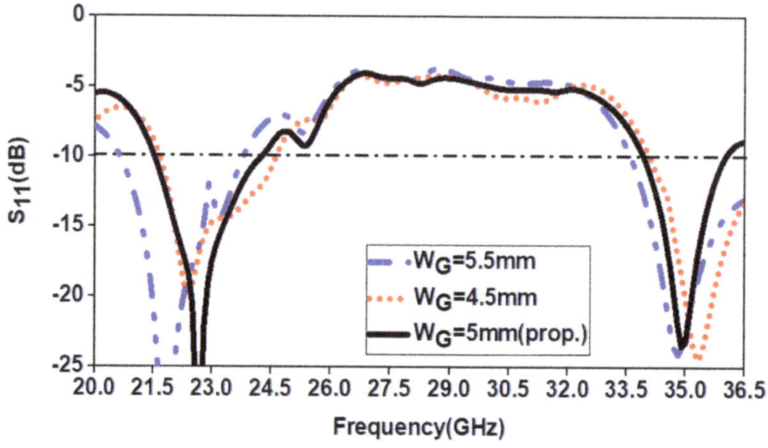

Figure 9. Analysis of variation in the width of first type ground cut W_G.

3.4. Effect of the Width of the Second Type of Ground Cut W_C

The effect of the width of the second type of ground cut on the antenna is analyzed. The width is varied to observe its impact. The variation in the width causes a significant variation in both the return loss and frequency at the first resonant frequency. In contrast, at the second frequency, the return loss and frequency changes are small. The variation of the width of the second type of ground cut W_C is shown in Figure 10.

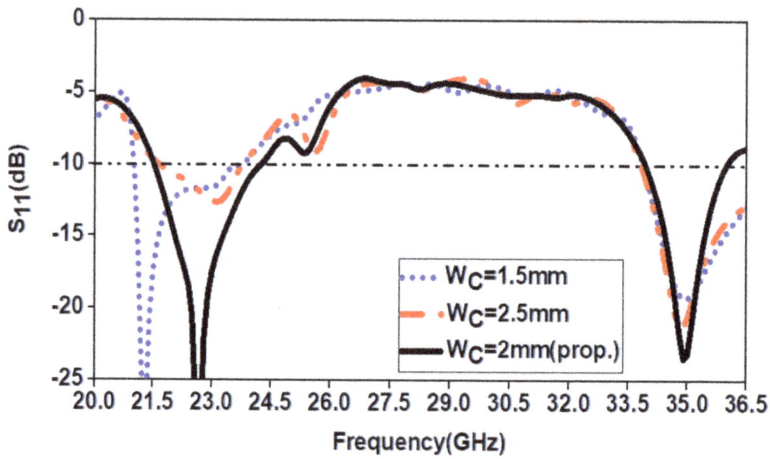

Figure 10. Analysis of variation in the width of second type ground cut W_C.

3.5. Effect of Length of First Type Ground Cut L_G

The effect of the width of the first type of ground cut on the antenna is studied. The width is varied to study its impact. The increase in the length results in variation in return loss at both frequencies, whereas the resonant frequency shifts at the first resonant value. The decrease in the length does not have a significant effect on the first resonant frequency. Still, at the second resonant frequency, both the frequency and return loss values are significantly changed. The variation of the length of the first type of ground cut L_G is given in Figure 11.

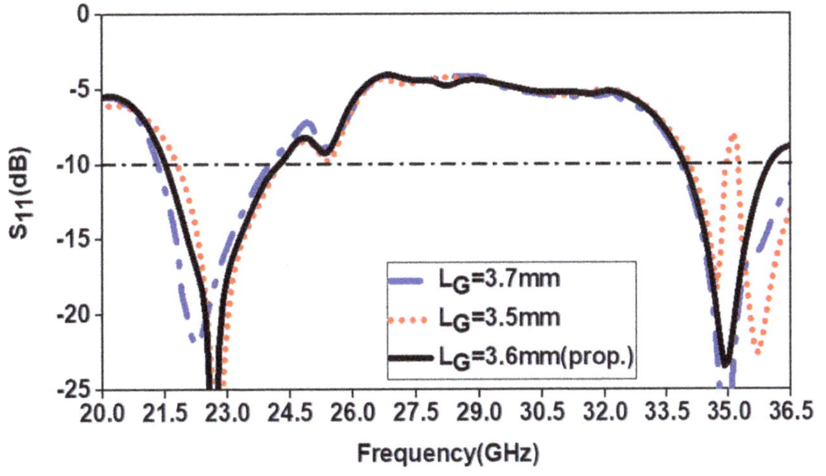

Figure 11. Analysis of variation in the length of first type ground cut L_G.

3.6. Effect of Length of Second Type Ground Cut L_C

The effect of the width of the second type of ground cut on the antenna is analyzed. The length is varied to observe its impact. The variation in the length does not significantly affect the frequency, but it results in a considerable variation in the return loss values at both resonant frequencies. The variation of the length of the second type of ground cut L_C is given in Figure 12.

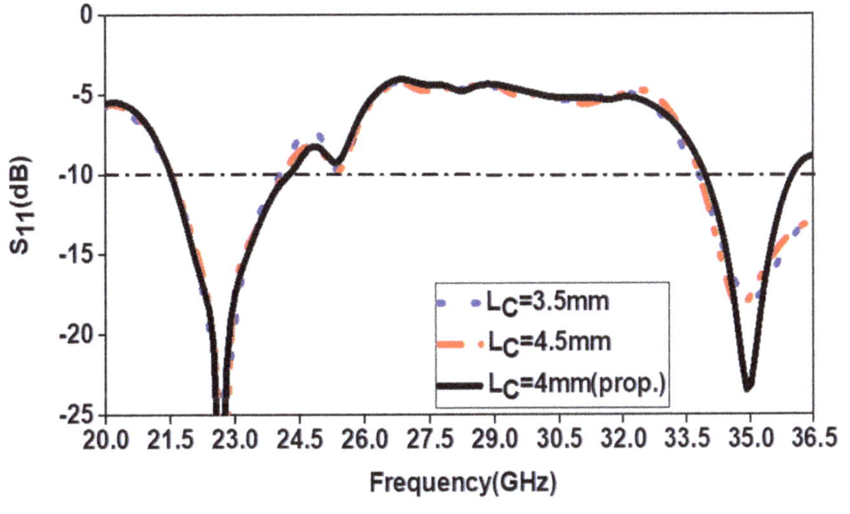

Figure 12. Analysis of variation in the length of second type ground cut L_C.

4. Results and Discussion

In the proposed design, the primary array antenna consists of a modified Franklin array structure printed on a substrate, which is of the material Rogers RT/duroid 6010, having a relative permittivity of 10.2 and a loss tangent of 0.0023. The simulation and measurement results for the S_{11} parameter vs. frequency of the antenna is depicted in Figure 13. The S_{11} result of the proposed antenna is measured using Anritsu 37269 A Vector Network Analyzer. It can be seen from the figure that the resonant frequency occurs for two values of operating frequency; 22.7 and 34.9 GHz in simulation, and at

22.8 and 35.3 GHz in measurement. In the simulation, for the first resonant frequency (i.e., 22.7 GHz), a bandwidth (for $S_{11} < -10$ dB criteria) of 2.8 GHz (21.5–24.3 GHz) and fractional bandwidth of 12.25% with maximum S_{11} of -30 dB is observed, and for the second resonant frequency (i.e., 34.9 GHz), a bandwidth of 2.1 GHz (33.9–36 GHz) and fractional bandwidth of 6% with maximum S_{11} of -23 dB is obtained. During measurement, for the first resonant frequency (i.e., 22.8 GHz), a bandwidth of 3 GHz (21.5–24.5 GHz) and fractional bandwidth of 13% with maximum S_{11} of -32 dB is observed, and for the second resonant frequency (i.e., 35.3 GHz), a bandwidth of 1.9 GHz (34.3–36.2 GHz) and fractional bandwidth of 5.3% with maximum S_{11} of -22 dB is obtained. Figure 13 clearly shows that the proposed structure achieves good impedance matching at both bands and the simulated and measured results are in close agreement.

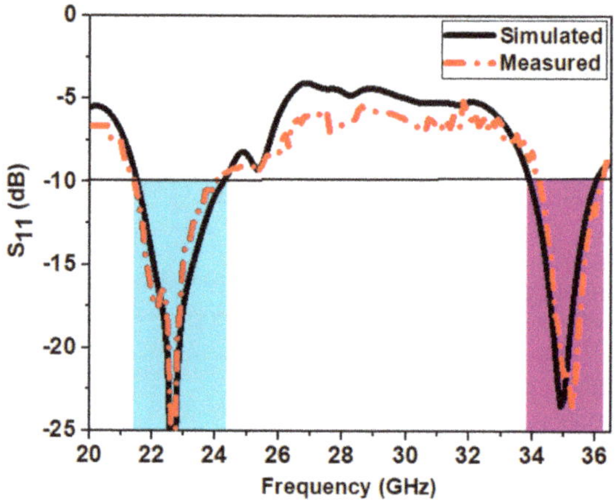

Figure 13. Reflection coefficient vs. frequency of the proposed antenna.

The surface current distribution patterns for the proposed antenna are analyzed in Figure 14 for both resonant frequencies. At 22.7 GHz, the maximum currents can be observed at the second and third arm from the left, while the corner of the ground creates more significant discontinuity to the flow of current at this band, as demonstrated in Figure 14a. At 34.9 GHz, the current leakage is more from the antenna's first and third arm from the left. Simultaneously, the slots in the ground allow more discontinuity in the current flow, thereby increasing the total current length path covering this particular band, as demonstrated in Figure 14b.

The 3D total gain pattern for the proposed antenna at both the resonant frequencies is shown in Figure 15. It is observed that the desired configuration of the antenna produces a maximum gain of 7.8 dB for 22.7 GHz and a gain of 9.7dB for 34.9 GHz. The gains are reasonably good as the bandwidth obtained at this band is relatively wide.

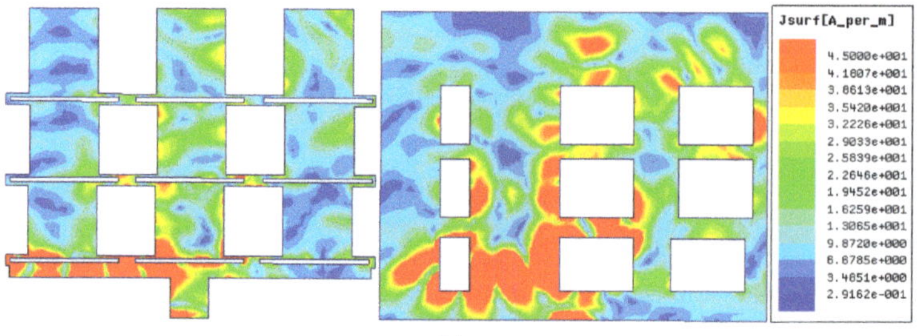

(a)

Figure 14. *Cont.*

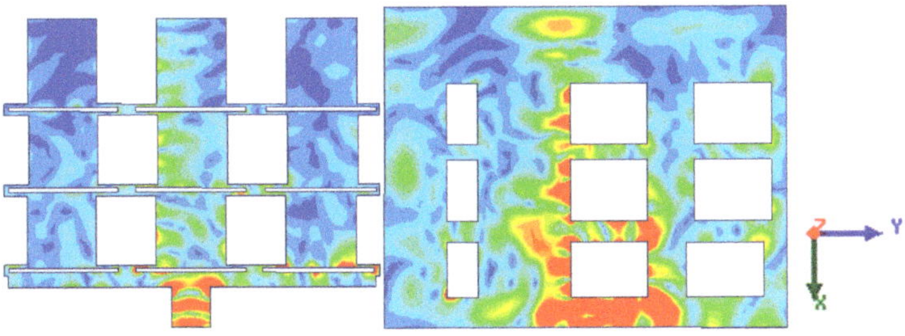

(**b**)

Figure 14. Surface current distribution plots for frequencies (**a**) 22.7 and (**b**) 34.9 GHz.

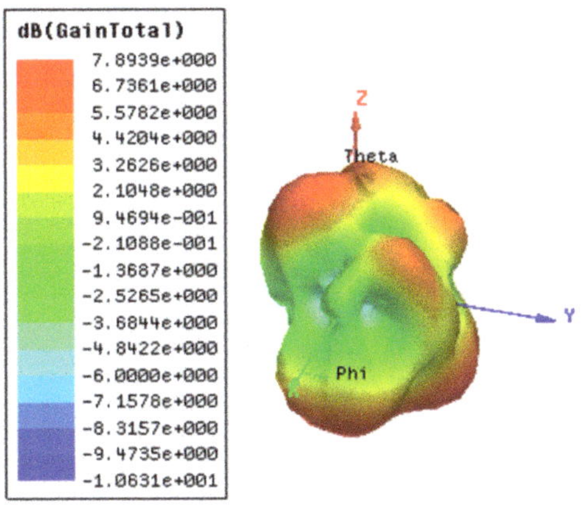

(**a**)

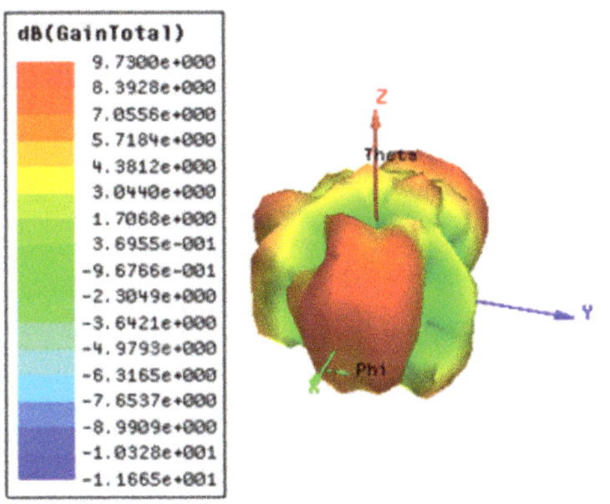

(**b**)

Figure 15. Three-dimensional gain plot for frequencies (**a**) 22.7 GHz (**b**) 34.9 GHz.

The variation of gain with frequency for the bandwidth of operation for the two resonant frequencies is analyzed, as shown in Figure 16. At the first band, i.e., 22.7 GHz, the gain is nonlinear and may be due to the dimension tuning of slots at ground plane to obtain the desired band. However, at the second band, i.e., 34.9 GHz, the gain varies linearly over the entire resonant bandwidth. The variation of radiation efficiency over the entire operating bandwidth for the two operating bands (i.e., 21.5–24.3 GHz, and 33.9–36 GHz) is also depicted in Figure 16c,d, respectively. It can be noticed that the antenna has radiation efficiency greater than 90%, at both the operating bandwidth, thus ensuring effective radiated power.

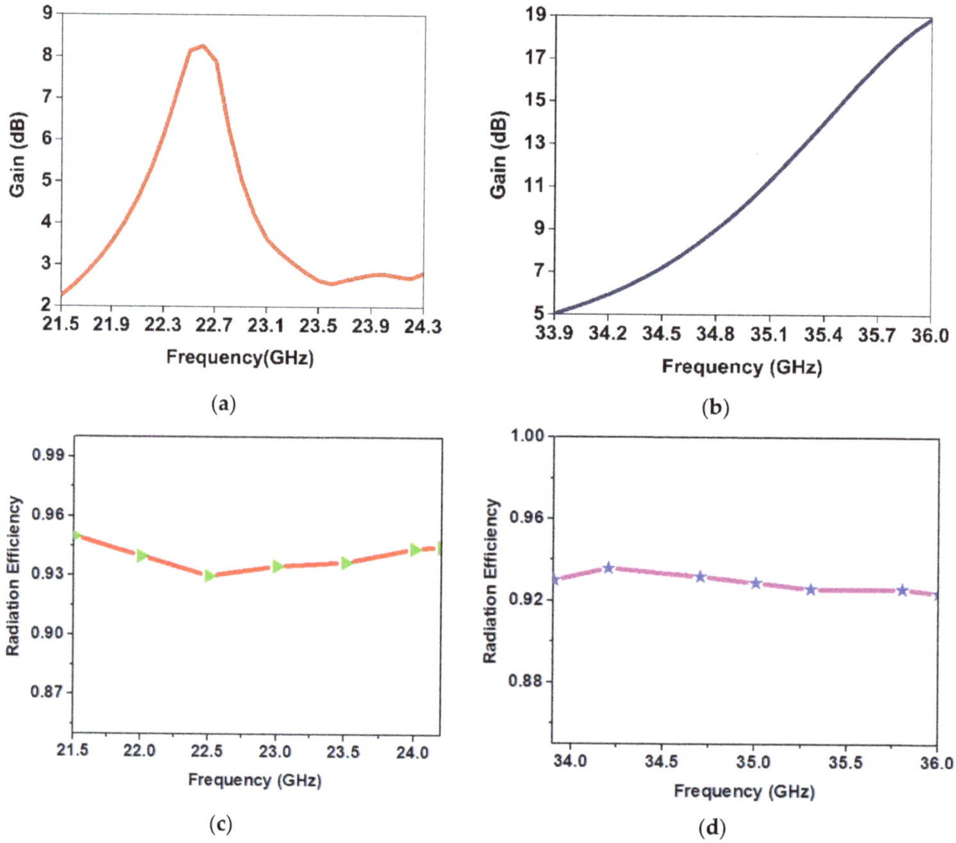

Figure 16. Variation of gain vs. frequency for the bandwidth of operation for a resonant frequency of (**a**) 22.7 and (**b**) 34.9 GHz, and variation of radiation efficiency vs. frequency for the bandwidth of operation for a resonant frequency of (**c**) 22.7 and (**d**) 34.9 GHz.

The radiation pattern for the proposed antenna for $Phi = 0°$ and $Phi = 90°$ is shown in Figure 17. It can be noted that the obtained pattern is directional and stable for both of the dual frequencies.

A comparison between the features of the proposed antenna with similar antennas is illustrated in Table 2. From the table, it can be found that the proposed structure has the advantage of compact size, multiband operation, directive radiation pattern, and a simple and planar structure over its counterparts.

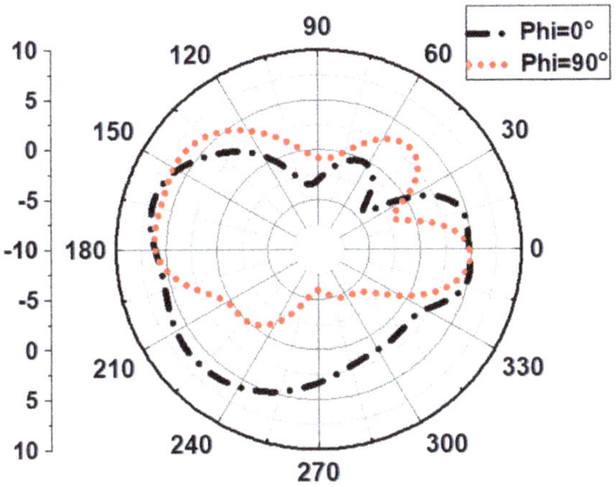

(**a**)

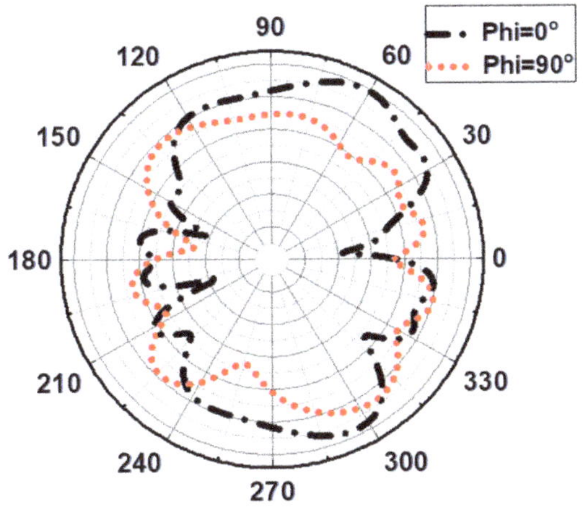

(**b**)

Figure 17. The radiation pattern for the proposed antenna at frequencies of (**a**) 22.7 and (**b**) 34.9 GHz.

Table 2. Comparison of various antenna performance indicators.

Ref.	Antenna Type	No of Elements	Resonating Frequency (GHz)	Bandwidth (%)	Gain (dBi)	Size (mm^2)	Advantage
[12]	Collinear Microstrip Patch Antenna	3	0.869	2.80	12.4	298 × 680	It has a directional pattern.
[13]	Collinear Microstrip Patch Antenna	3	2.4	7.10	15.8	298 × 680	Simple structure without the need for any feeding network.
[14]	Modified Microstrip Franklin Array Antenna	1, 3, and 5	24	1.04	6, 8, and 11	90 × 25	High absolute gain, good directivity, and large-range coverage.
[15]	Franklin Array Antenna	32	2.4	8.33	12	520 × 520	High gain, enlarged antenna aperture, increase in available power.

Table 2. *Cont.*

Ref.	Antenna Type	No of Elements	Resonating Frequency (GHz)	Bandwidth (%)	Gain (dBi)	Size (mm^2)	Advantage
[17]	Circularly Polarized Franklin Microstrip Antenna	12	6.9	5.76	3.1	172 × 172	More than 64% of the −10 dB impedance bandwidth is usable in CP mode, low axial ratio, high inter-port isolation, and squinted beams.
[18]	Conformal Antenna array	6	29	19.78	8.3	12 × 28.2	Compact, flexible, and wideband antenna array
[19]	Dielectric Slab Antenna based on microstrip-Franklin Excitation	1	30	7.6	6.6	80 × 20	Simple planar geometry.
[20]	Enhanced Franklin antenna with defected ground structure	9	25.2	22.22	10.1	26 × 21	Directive radiation pattern and cost effective
	Proposed Antenna	9	22.7 and 34.9	12.2 and 6	7.8 and 9.73	26 × 21	Compact size, multiband operation, directive radiation pattern, simple and planar structure.

5. Conclusions

A modified Franklin antenna array with a slotted ground structure to be utilized for mmW frequency is proposed. The nine patch antennas and folded dipole-like structures arranged in a three-by-three array manner are connected to a feeding network modified into a folded dipole like structure to provide good impedance matching to the overall antenna design. To achieve the optimum dimensions and suitable results, parametric analysis was carried out by modifying the antenna's various parameters. The antenna offers dual-band resonance characteristics at 22.7 and 34.9 GHz with a total impedance bandwidth of about 2.8 and 2.1 GHz in simulation. For measurement, it shows dual-band operation at 22.8 and 35.3 GHz with impedance bandwidth of about 3 and 1.9 GHz, respectively, thus showing a good correlation between the simulated and measured results. The concept of the antenna array is deployed to enhance the gain of the antenna at a higher frequency of operation and nullify the side lobes. It also provides better impedance matching, as well as providing diversity reception. The antenna offers good gain, impedance matching, compact size, and stable radiation pattern, and thus acts as a very competitive candidate for 5G applications.

Author Contributions: All authors contributed equally for this work. All authors have read and agreed to the published version of the manuscript.

Funding: This research received no external funding.

Institutional Review Board Statement: Not applicable.

Informed Consent Statement: Not applicable.

Data Availability Statement: Not applicable.

Conflicts of Interest: The authors declare no conflict of interest.

References

1. Goldsmith, A. *Wireless Communications*; Cambridge University Press: New York, NY, USA, 2005.
2. Balanis, C.A. *Antenna Theory: Analysis and Design*; John Wiley & Sons: New York, NY, USA, 2016.
3. Huang, K.-C.; Zhao, C. *Millimetre Wave Communication Systems*; Wiley: Hoboken, NJ, USA, 2011.
4. Ali, T.; Aw, M.S.; Biradar, R.C. A Compact Bandwidth Enhanced Antenna Loaded with SRR For WLAN/WiMAX/Satellite Applications. *Adv. Electromagn.* **2018**, *7*, 78–84. [CrossRef]
5. Ali, T.; Subhash, B.K.; Pathan, S.; Biradar, R.C. A compact decagonal-shaped UWB monopole planar antenna with truncated ground plane. *Microw. Opt. Technol. Lett.* **2018**, *60*, 2937–2944. [CrossRef]
6. Ndip, I.; Le, T.H.; Schwanitz, O.; Lang, K.-D. A comparative analysis of 5G mmWave antenna arrays on different substrate technolgies. In Proceedings of the 2018 22nd International Microwave and Radar Conference (MIKON), Poznan, Poland, 14–17 May 2018; pp. 222–225.
7. Jilani, S.F.; Alomainy, A. A multi-band millimeter-wave 2-D array based on enhanced Franklin antenna for 5G wireless systems. *IEEE Antennas Wirel. Propag. Lett.* **2017**, *16*, 2983–2986. [CrossRef]
8. Pi, Z.; Khan, F. An introduction to millimeter-wave mobile broadband systems. *IEEE Commun. Mag.* **2011**, *49*, 101–107. [CrossRef]
9. Solbach, K. Microstrip-Franklin Antenna. *IRE Trans. Antennas Propag.* **1982**, *30*, 773–775. [CrossRef]
10. Alrifai, A.A. Improving the Frequency Range of Franklin's Antenna. *Int. J. Commun. Antenna Propag. (IRECAP)* **2017**, *7*, 410. [CrossRef]
11. Maharjan, J.; Kim, S.-W.; Choi, D.-Y. Franklin Array MIMO Antenna for 5G Applications. In Proceedings of the 2019 34th International Technical Conference on Circuits/Systems, Computers and Communications (ITC-CSCC), JeJu, Korea, 23–26 June 2019; pp. 1–4.
12. Polivka, M.; Holub, A.; Mazanek, M. Collinear microstrip patch antenna. *Radioengineering-prague* **2005**, *14*, 40.
13. Holub, A.; Polivka, M. Collinear microstrip patch antenna. In *Passive Microwave Component and Antennas*; InTech: Rijeka, Croatia, 2010; pp. 513–530.
14. Kuo, C.H.; Lin, C.C.; Sun, J.S. Modified microstrip Franklin array antenna for automotive short-range radar application in blind spot information system. *IEEE Antennas Wirel. Propag. Lett.* **2017**, *16*, 1731–1734. [CrossRef]
15. Chang, S.-H.; Liao, W.-J.; Peng, K.-W.; Hsieh, C.-Y. A Franklin Array Antenna for Wireless Charging Applications. *PIERS Online* **2010**, *6*, 340–344. [CrossRef]
16. Wang, P.P.; Antoniades, M.A.; Eleftheriades, G.V. An Investigation of Printed Franklin Antennas at X-Band Using Artificial (Metamaterial) Phase-Shifting Lines. *IEEE Trans. Antennas Propag.* **2008**, *56*, 3118–3128. [CrossRef]
17. Chaudhuri, S.; Kshetrimayum, R.S.; Sonkar, R.K. High Inter-port Isolation Dual Circularly Polarized Modified Franklin Microstrip Antenna. In Proceedings of the 2019 13th European Conference on Antennas and Propagation (EuCAP), Krakow, Poland, 31 March–5 April 2019; pp. 1–4.
18. Jilani, S.F.; Alomainy, A. Millimeter-wave conformal antenna array for 5G wireless applications. In Proceedings of the 2017 IEEE International Symposium on Antennas and Propagation & USNC/URSI National Radio Science Meeting, San Diego, CA, USA, 9–14 July 2017; pp. 1439–1440.
19. Boas, E.C.; Alves, A.A.C.; Ribeiro, J.A.J.; Cerqueira, S., Jr. A Novel Dielectric Slab Antenna Based on Microstrip-Franklin Excitation for mm-Waves. Journal of Microwaves. *Optoelectron. Electromagn. Appl.* **2020**, *19*, 203–213. [CrossRef]
20. Vanaja, C.; Pavithra, N.; Sravya, N.; Manoj, M.; Dhanade, Y.B. Enhanced Franklin Antenna for the Future 5G Communication Applications. *IJITEE* **2019**, *8*, 1–5.

applied sciences

MDPI

Article

Design of SWB Antenna with Triple Band Notch Characteristics for Multipurpose Wireless Applications

Warsha Balani [1], Mrinal Sarvagya [1], Tanweer Ali [2,*], Ajit Samasgikar [3], Saumya Das [4], Pradeep Kumar [5,*] and Jaume Anguera [6,7]

1 School of Electronics and Communication Engineering, Reva University, Bangalore 560064, India; balani.warsha@gmail.com (W.B.); mrinalsarvagya@reva.edu.in (M.S.)
2 Department of Electronics and Communication, Manipal Institute of Technology, Manipal Academy of Higher Education, Manipal, Karnataka 576104, India
3 MMRFIC Technology Pvt Ltd., Bangalore 560016, India; ajitsamasgikar@gmail.com
4 Department of Information Technology, Sikkim Manipal Institute of Technology, Sikkim Manipal University, Sikkim 737102, India; saumya.das@smit.smu.edu.in
5 Discipline of Electrical, Electronic and Computer Engineering, University of KwaZulu-Natal, Durban 4041, South Africa
6 Fractus Antennas, 08174 Barcelona, Spain; jaume.anguera@salle.url.edu
7 Electronics and Telecommunication Department, Universitat Ramon LLull, 08022 Barcelona, Spain
* Correspondence: tanweer.ali@manipal.edu (T.A.); kumarp@ukzn.ac.za (P.K.)

Abstract: A compact concentric structured monopole antenna for super wide band (SWB) applications with triple notch band characteristics is designed and experimentally validated. The antenna covers an immense impedance bandwidth (1.6–47.5 GHz) with sharp triple notch bands at 1.8–2.2 GHz, 4–7.2 GHz, and 9.8–10.4 GHz to eliminate interference from co-existing advanced wireless services (AWS), C bands, and X bands, respectively. By loading an E-shaped stub connected at the top of the patch and by etching a split elliptical slot at the lower end of the radiating patch, the band rejection characteristics from 1.8–2.2 GHz for the AWS and 4–7.2 GHz for the C band are achieved, respectively. Further, by making use of a C-shaped resonator near the feed line, band rejection from 9.8–10.4 GHz for the X band is obtained. By varying the parameters of the antenna, the notch bands are controlled independently over a wide range of frequencies. The antenna provides good radiation characteristics, constant group delay response, and better gain over the pass band. The experimental results indicate that the designed antenna offers a remarkable reduction in gain and high variation in group delay over the stop bands. To characterize the wideband property and linear phase response of the designed antenna, its time-domain performance is extensively described and evaluated, which assure pulse transmission with minimum distortion.

Keywords: super wide band antenna (SWB antenna); triple band notch; time-domain characterization

Citation: Balani, W.; Sarvagya, M.; Ali, T.; Samasgikar, A.; Das, S.; Kumar, P.; Anguera, J. Design of SWB Antenna with Triple Band Notch Characteristics for Multipurpose Wireless Applications. *Appl. Sci.* **2021**, *11*, 711. https://doi.org/10.3390/app11020711

Received: 8 December 2020
Accepted: 4 January 2021
Published: 13 January 2021

1. Introduction

1.1. Motivation

In recent years, wireless communication systems have grown at a significant rate. Wireless consoles for future applications are necessary to support distinct services. This increasing demand leads to the requirement for antennas that are able to cover multiple bands for accommodating several communication standards within a single system in consideration of the fact that the frequency spectrum allocated for ultrawideband (UWB) by the Federal Communications Commission (FCC) is from 3.1 to 10.6 GHz. There is a requirement for a competent super wide band (SWB) antenna that should be capable of functioning over the entire ultra-wide band, as assigned by the FCC, and that should cover lower frequency bands. However, the SWB system broad frequency range may influence the present narrowband systems. Therefore, to alleviate the electromagnetic interference effect of an SWB system over the narrowband system and vice-versa, an SWB antenna with a band-notch feature is desirable for coexisting wireless communications.

A conventional approach for preventing intervention among the recently designed and present-day communication systems is to insert filter banks in the SWB system to filter out the interfering EM waves of different frequencies. However, introducing external band stop filters to accomplish the desired band rejection makes the system complex and enhances the dimension significantly.

Another efficient way for suppressing interfering electromagnetic spectrums from the SWB system is to realize SWB antennas with frequency blocking ability. Antenna shape and structure orientation are two important aspects in designing this type of antenna structure. Several wideband antennas are reported in the literature with no notch, single notch, or dual notch functions. Nevertheless, there is a need to develop SWB antennas with a greater number of notch bands to establish multiple coexisting communications. However, it is very challenging to design an SWB antenna with multi notch bands due to the difficulty in bandwidth control of notch bands while maintaining space limitation and unwanted mutual coupling between the adjacent notched bands.

1.2. Literature Review

The major difficulty in implementing an SWB communication system is eliminating the authorized and unauthorized license bands such as WiMAX, WLAN, AWS, C, X, etc., from the assigned SWB bandwidth. Each of these communications has its defined frequency bands, and the SWB system has to separate out those bands to make the communication interference resistant.

Various researchers have addressed the aforementioned challenge to some extent by introducing SWB antennas with multi notch characteristics. Several techniques such as the introduction of a slot in the patch/ground plane [1–3]; insertion of a slit on the patch/ground plane [4,5]; placing stubs [6–9], meander grounded stubs [10], and parasitic elements [11–13]; and employing a split ring resonator [14,15] have been used in SWB antenna design to filter out undesired frequency bands.

Placing a slot on the patch or ground plane empirically helps to reject a particular frequency band and thus brings band notch characteristics to the SWB antenna. Various shapes and sizes of slots have been investigated, and combinations of slots with other geometries have been examined for rejecting a desired band of frequencies. In [1], the bandwidth of a square-shaped antenna was enhanced by placing a dumbbell-structured parasitic resonator at the upper side of the partial ground plane. Further, by adding two similar inverted J-shaped slots and a reverse F-shaped slot, two notch bands were obtained at 3.3–4.2 GHz and 5.1–5.4 GHz frequencies. Again, in [2], a triple notch band SWB antenna was designed by exploring the slot on patch concept. First, a notch band at 3.5 GHz (WiMAX) was realized by cutting an E-shaped slot in the upper portion of the antenna, while a rectangular slot and a C-shaped slot along with a set of two parasitic patches provided two more notch bands at 5.2 GHz and 5.8 GHz (lower and upper) WLAN, respectively. A monopole printed circular radiating patch with a coplanar waveguide (CPW)-fed bevel-shaped ground plane is demonstrated in [3]. By etching two C-shaped slots on the radiating patch and a pair of C-shaped slots in the ground plane, three frequency bands, namely 3.43–3.65 GHz, 4.95–5.25 GHz, and 5.36–5.85 GHz, were eliminated.

Introduction of slits on the patch results in multiband frequency operation because of the disruption caused to the average current paths of any resonant mode. The surface current paths become elaborated around the slits, and because of this, resonant frequency decreases leading to multiband frequency operation. In [4], a modified circular radiating patch with a pair of open ended quarter wavelength slits was realized to suppress dual notch bands of 5.15–5.35 GHz and 5.725–5.825 GHz. In another example of the slit concept, a CPW-fed single band notch circular fashioned antenna was demonstrated in [5]. With the help of a set of stubs and slits adopted inside the tapered slot and circular patch, a notch band of 5.0–5.8 GHz was realized.

Another method to bring notches in wideband antennas is by loading the stub on the antenna structure. Stub loading on the patch or ground plane transfuses the fundamental and harmonics in quadrature and thus eliminates the required band of frequencies. Jiang et al. presented a square-shaped radiating patch with a modified ground plane for developing a dual notch band wideband antenna. By introducing an inverted T-shaped stub on the patch and a set of U-shaped stubs adjacent to the feed line, stop bands of 3.3–4.0 GHz (WiMAX) and 5.05–5.90 GHz (WLAN) were achieved [6]. In [7], a quasi-U-shaped patch with a stepped slot in the ground plane was presented to bring wide bandwidth in antenna characteristics. Two inverted L-shaped stubs in the radiating patch realized two notch bands of 4.97–5.48 GHz and 5.69–5.99 GHz for removing lower and upper WLAN bands, respectively. A guitar-shaped CPW-fed monopole radiator with triple band notch functionality was presented by Sharma et al. [8]. By adding an inverted T-shaped stub and a pair of elliptical slots on the radiator, three interfering frequencies of 3.74 GHz (under WiMAX band), 5.45 GHz (under WLAN band), and 7.65 GHz (under X band) were eliminated. In [9], by embedding a fractal stub on the radiating patch and by etching a pair of modified rectangular slots, triple notch bands from WiMAX (3.30–3.80 GHz), WLAN (5.150–5.825 GHz), and DSS (7.25–7.75 GHz) were encountered.

Alternately, meander lines have also been used for notch antenna design, because they resonate at different independent frequencies. Meandering means turning or winding. If the number of turns are increased, the meander stub resonates at lower frequencies, and vice versa. It is well known fact that the surface current primarily distributes at the periphery of the microstrip line, and in the meander stubs, the magnitude of surface current will be greater at the corners/bending portion, which provides a coupling to the microstrip line, thus providing the notch at that particular frequency. In [10], a microstrip-fed elliptical-shaped planar monopole antenna with three notch band properties was demonstrated. With the aid of three sets of meandered ground stubs, triple notch band characteristics were obtained. Two notch bands of 3.3–3.6 GHz and 5.15–5.35 GHz were derived from two sets of meandered stubs near the feedline, and another notch band of 5.725–5.825 GHz was yielded from a meandered stub on the upper edge of the ground plane. Another technique to obtain notch bands in a wideband antenna is by introducing parasitic elements near the radiator. A parasitic element is an element that operates under the influence of other's feed. The parasitic element behaves like a passive resonator that modifies the radiation pattern emitted by the patch antenna. Thus, the parasitic element helps to provide wideband characteristics and notch functionality. Yadav et al. [11], presented a circular monopole radiator with a parasitic resonator for adding wideband characteristics. By embedding I-shaped and inverted-U-shaped parasitic resonators at the rear side of the radiating patch, dual band notch functions at 5–5.35 GHz and 7.85–8.4 GHz were achieved. Wide bandwidth was also achieved by introducing parasitic elements in addition to slot concepts in antenna design. A tapered triangular microstrip-fed circular monopole antenna with dual band notch characteristics was demonstrated by exploring the parasitic element and slot concept in [12]. By inserting a U-shaped parasitic component in proximity of the ground plane and an inverted T-shaped stub inscribed within the patch, a first stop band at 3.2–4.4 GHz was obtained. For the second stop band at 7.2–8.4 GHz, a U-shaped slot was etched out from the ground plane. Another illustration of a parasitic element-based wideband antenna with band notch characteristics can be seen in [13], where a hemi ellipse-shaped radiating patch with a parasitic element and an elliptical slot was used to hold a single notch band (5.15–5.825 GHz) in the complete frequency band of 3–17 GHz.

In another approach, a split ring resonator (SRR) and complementary SRR (CSRR) have been significantly used for designing antenna with band notch capability. The SRR and CSRR can produce strong electric and magnetic coupling from LC resonance and thus are used for band stop filter modelling. In [14], by etching two CSRRs on the radiating patch and by using a set of rectangular SRRs beside the feed line, three notch bands of 3.3–3.8 GHz (Wi-MAX), 5.15–5.85 GHz (WLAN), and 7.9–8.4 GHz (X-band) were brought in the complete frequency band of 2.21–11.71 GHz. In another example of a CSRR-based wideband antenna, a miniaturized microstrip-fed planar monopole antenna with three band notch features was designed by Sarkar et al. [15]. There, triple band notch (3.4 GHz, 5.2 GHz, and 5.8 GHz) functionality was achieved by defining a set of slots around the edges of the patch, a circular-shaped CSRR at the center of the patch, and a set of rectangular-shaped CSRRs on either side of the ground.

The techniques presented in the foregoing discussion are able to create single or multiple notch bands for wideband planar antennas. It has been observed that most of these wideband antennas are able to accommodate frequencies up to 20 GHz, but present-day requirements of many wireless communications such as Ka band satellite, 5G, 6G, etc., are beyond this range. Therefore, there is a need to design an SWB antenna that can support communication beyond 20 GHz. At the same time, the SWB antenna needs to be interference resistant for establishing coexisting communication. Therefore, in addition to increasing bandwidth for the SWB antenna, multiple notch characteristics need to be implemented to avoid interference from licensed and unlicensed existing communications.

Another major concern that has not been addressed by the majority of existing research on wideband antennas is the time-domain characterization that includes measurements of group delay, impulse response, system fidelity factor, etc. Time-domain characterization of the SWB antenna has a greater significance, as those antennas exploit a severe influence on the waveforms.

1.3. Contributions

In this work, design and analysis of an SWB antenna with triple band notch characteristics for multiple wireless applications are presented. The designed antenna is capable of removing three interfering bands at advanced wireless services (AWS), and C and X band systems.

- To attain the notched frequency band for AWS (1.8–2.2 GHz), an E-shaped stub is loaded on the top of the radiator.
- A split elliptical slot at the base of the radiating patch is used to achieve a band notch at the C band (4–7.2 GHz).
- By placing a C-shaped resonator near the feed, a band notch at the X band (9.8–10.4 GHz) is acquired.

The main prominence of the designed radiator is its very wide bandwidth and large bandwidth dimension ratio (BDR), which is obtained by modifying the electrical length of the radiator. The presented antenna structure attains a frequency range of 1.67–47.5 GHz (VSWR < 2) with a bandwidth ratio of 28.44:1 and a fractional bandwidth of 186%. The results are captured by simulating the model in HFSS 19.0 software.

This paper is organized into different sections as follows: The first section presents an introduction to the band-notched SWB antenna. The design steps and characteristics of the proposed antenna are investigated in Section 2. Significant parameters that control the bandwidth and impedance matching are illustrated in Section 3. Then the simulated and experimental results viz. current distribution, gain, and radiation patterns are illustrated in Section 4. Section 5 focuses on modelling of equivalent circuit for the proposed antenna. Time-domain characteristics, namely system fidelity factor (SFF), group delay, and transfer function, are presented in Section 6. Section 7 highlights the main features of the proposed SWB antenna by including a comparative analysis. Finally, the paper is concluded in Section 8.

2. Design Methodology

The geometry of the proposed SWB monopole antenna with triple band-notched characteristics is depicted in Figure 1, its related parameters are indexed in Table 1, and the configuration is detailed analysis has been given in [16]. The antenna was implemented on a Rogers RT-Duroid 5880 substrate with dielectric constant of 2.2, tan δ = 0.0009, and thickness of 1.57 mm. A procedural evaluation of the presented antenna was carried out in order to analyze the behavior of the designed antenna at particular steps. The main purpose of describing the present section is to explain the technique for obtaining three notched interfering bands at the coinciding frequencies. This section also clarifies the outcome of each notch-creating configuration on the functioning of the reference antenna. Each individual stage illustrates the characteristics of the antenna in terms of VSWR.

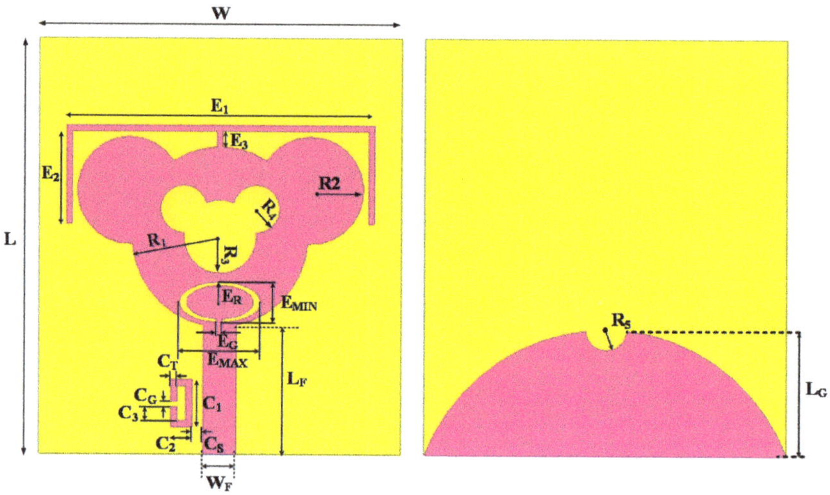

Figure 1. Labelled geometry of the triple band-notched super wide band (SWB) antenna front view and back view made on RT-Duroid 5880 substrate with dielectric constant of 2.2, tan δ = 0.0009, and thickness of 1.57 mm.

Table 1. Optimal values for different parameters of the designed triple band-notched SWB antenna structure.

Symbol	Dimension (mm)	Symbol	Dimension (mm)
W	40	R4	2.6
L	45	R5	2.3
R1	9.7	L_F	21.4
R_2	5.8	W_F	3.6
R_3	4	L_G	15
E_1	34	C1	5
E_2	10	C2	2.4
E_3	1.2	C3	1.6
E_{MIN}	4.2	CG	0.6
E_{MAX}	8.16	CT	0.54
E_G	0.6	CS	0.6
E_R	0.2		

In phase 1, a microstrip-fed structural monopole antenna for SWB application is examined. In the second phase, design and exploration of the SWB antenna with a single notch-band is exhibited. In the third phase, a printed SWB antenna is described with dual-band notch characteristics at two unique interference bands. In the final phase, a planar SWB antenna with triple band notch properties is introduced. The notch bands can be independently tuned to reject individual frequency bands.

2.1. Phase 1: Design of Super Wideband Antenna (Antenna 1)

A microstrip-fed printed monopole antenna for SWB application is used. The proposed antenna is derived from a conventional circular monopole antenna. To achieve wide bandwidth and matching, a conventional circular monopole antenna was modified by adjoining a pair of ears at the upper part of the radiator and also by modifying its partial ground plane. The anatomy of the designed SWB antenna and its corresponding VSWR variation over frequency are displayed in Figure 2. The values of VSWR between 1 and 2 throughout the frequency range (1.2–47.5 GHz) depicts good impedance matching for the proposed SWB antenna. Since there is no predefined standard frequency spectrum for SWB antenna structures, a comparison between the wideband antenna structures can be made in terms of an index term, i.e., the bandwidth dimension ratio (BDR) [16], which defines the antenna bandwidth (in percentage) per unit area of the antenna. The following equation is used for BDR calculation:

$$BDR = \frac{BW\%}{\lambda_{length} \times \lambda_{width}} \tag{1}$$

where, λ_{width} and λ_{length} are the electrical width and length, respectively, of the antenna in the context of wavelength associated with the lower cut-off frequency, and $BW\%$ is percentage bandwidth considering a VSWR ≤ 2. The designed antenna structure achieved a total bandwidth range of 1.2–47.5 GHz (VSWR < 2) along with a *BDR* of 38.9:1 and a fractional bandwidth of 190%.

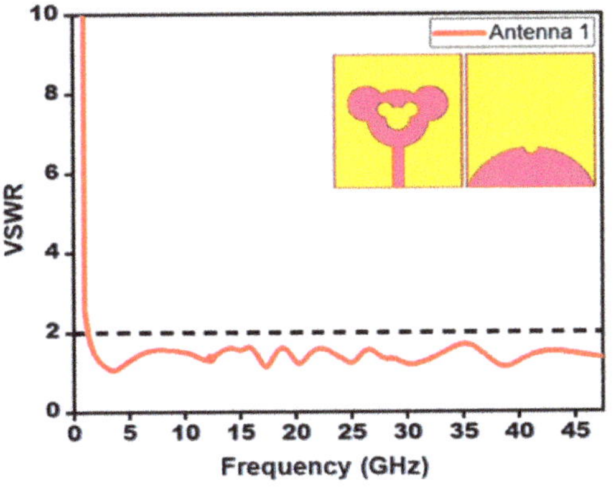

Figure 2. Phase 1: VSWR result of the SWB antenna.

2.2. *Phase 2: Design of Single Notched Band SWB Antenna (Antenna 2)*

The most critical concern in the design of a band-notched SWB antenna is the deployment of its notch-forming configurations to prevent nearby existing bands. The notch-band centered at 2 GHz (AWS), covering the frequency range of 1.8–2.2 GHz was prevented by placing an E-shaped stub on the top of the radiator. The antenna with a single band notch feature is presented in Figure 3. It can be noted that the dimensions of the reference designed SWB antenna (Antenna 1) remained unaltered when the notch-creating configuration was included with the reference antenna. Therefore, no re-tuning was needed for the reference SWB antenna for introducing the notch characteristics. The simulated VSWR of single-notched band SWB antenna, as shown in Figure 3, demonstrated the rejection of the frequency band of 1.8–2.2 GHz from the entire band of 1.2–47.5 GHz.

2.3. *Phase 3: Design of Dual Band-Notched SWB Antenna (Antenna 3)*

To deal with undesirable possible intervening of the narrowband regime with the SWB antenna, another intervention band in extension with the AWS band was created at 6 GHz, which was capable of covering the bandwidth of 4–7.2 GHz (C band). A split elliptical-shaped slot was engraved at the lower edge of the patch to stop the C band. Figure 4 represents the antenna structure with two band-notches and its VSWR result. From the VSWR graph, we can deduce that the second band notch introduced in the antenna structure did not affect the efficacy of the single band notch (Antenna 2).

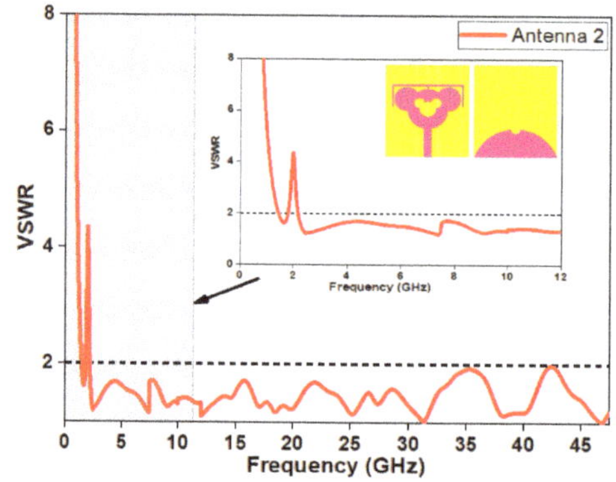

Figure 3. Phase 2: VSWR of the single band-notched antenna.

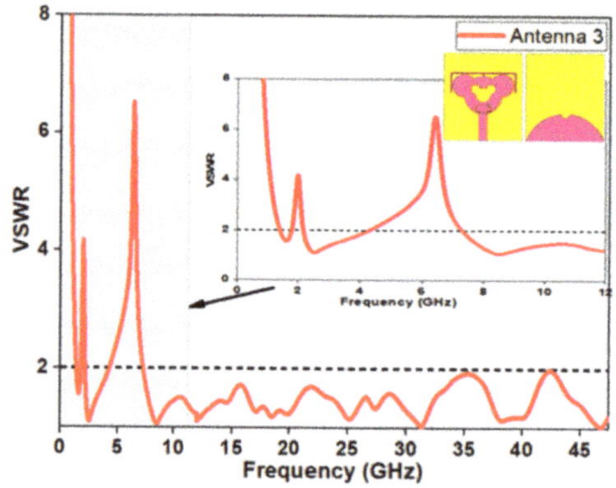

Figure 4. Phase 3: VSWR of the dual band-notched antenna.

2.4. Phase 4: Design of Triple Band-Notched SWB Antenna (Proposed Antenna)

To alleviate the interference at X band, a C-shaped resonator was placed at the bottom of the patch beside the feedline. The addition of the C-shaped resonator led to the obstruction of the X band (9.8–10.4 GHz) without influencing the efficacy of Antenna 3. Figure 5 illustrates the proposed antenna configuration and VSWR result of the triple band-notched antenna. The designed antenna exhibited a wide bandwidth, including a triple notch band feature with a very compact design.

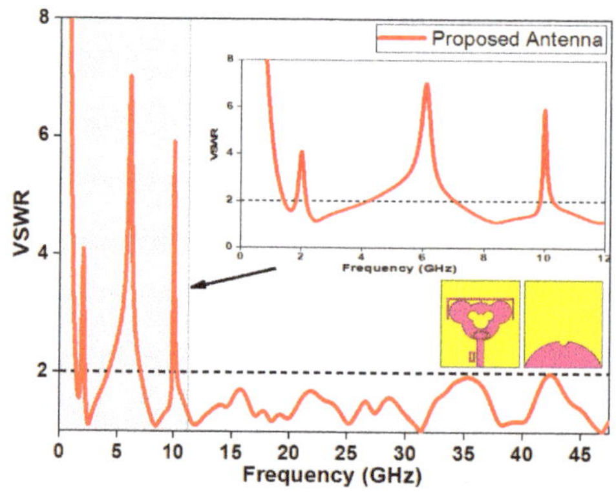

Figure 5. Phase 4: VSWR result of the triple band-notched antenna.

3. Parametric Analysis

A parametric study was conducted for optimizing numerous design parameters. A comprehensive behavioral analysis of the designed antenna was accomplished with a view to evaluating the efficacy of the antenna and to identify the influencing parameters of the respective notch band. The impact of the notching factors on the respective notches was investigated thoroughly, whereas the rest of the factors remained unaltered.

3.1. Independently Controllable Notch AWS Band (1.8–2.2 GHz)

Figure 6 illustrates the optimization of an "E"-shaped stub accountable for mitigating AWS (1.8–2.2 GHz) interference. It is evident from Figure 6 that E2 was the controlling parameter of the notch for the AWS band of the proposed antenna.

$$L_{stub} = \frac{c}{2f_{notch} \cdot \sqrt{\varepsilon_{eff}}} \tag{2}$$

$$\varepsilon_{eff} = \frac{\varepsilon_r + 1}{2} \tag{3}$$

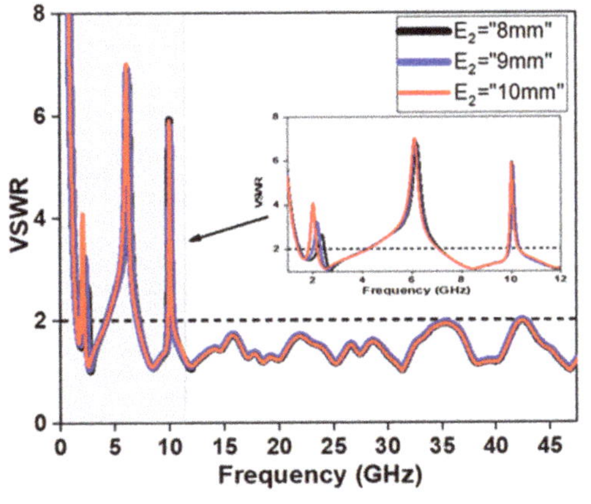

Figure 6. Parametric variation of independently controllable AWS band notch by varying E2.

For the optimized dimension of E2 = 10 mm, the first frequency band notch was centered at 2 GHz, the second frequency band notch was centered at 6 GHz, and the third frequency notch was centered at 10 GHz. When the parameter E2 was modified from 8 mm to 9 mm, the first notch band shifted in the regime of the higher frequency range, whereas the second and third notch bands remained unaltered. The optimized equation of the "E"-shaped stub affecting the AWS wireless communication system is given by Equation (2), where (E1 = 34 mm, E2 = 10 mm, E3 = 1.2 mm) is the stub length, ε_r is relative permittivity, ε_{eff} is effective dielectric constant of the substrate calculated using Equation (3), "c" corresponds to the speed of light, and f_{notch} is the notch center frequency. The theoretically calculated "E"-shaped stub length for 2 GHz is 59.29 mm, while the optimum value for the "E"-shaped stub is 55.2 mm. The detailed parametric variations is illustrated in Table 2.

Table 2. Parametric variation of notched bands.

AWS		C Band		X Band	
E_2 (mm)	Bandwidth (GHz)	E_{MIN} (mm)	Bandwidth (GHz)	C_2 (mm)	Bandwidth (GHz)
8	2.1–2.5	4	4.11–7.14	2.2	10.2 to 10.69
9	2–2.4	4.2	4–7.13	2.3	10.06 to 10.57
10	1.8–2.2	4.4	3.79–7.05	2.4	9.86 to 10.34

3.2. Independently Controllable Notch C Band (4–7.2 GHz)

The elliptical slot at the lower radiating portion of the designed antenna was liable for notch creation at the C band. It is apparent from the Figure 7 that parameter EMIN was accountable for governing the notch at the C band. For EMIN = 4 mm, the notch at the C band was at 6.13 GHz. By increasing the value of EMIN from 4 mm to 4.2 mm, the second notch was shifted from 6.13 GHz to 6 GHz. By further increasing the value of EMIN to 4.4 mm, the second notch was reached at 5.97 GHz. In all instances, the first and second notches stayed unaffected by changes in the factor EMIN. It could be stated from the study of the parametric analysis for EMIN that the middle notch could be independently controlled. It was also ascertained that the EMIN parameter did not influence the first frequency notch band (AWS band) or the third frequency notch band (X band).

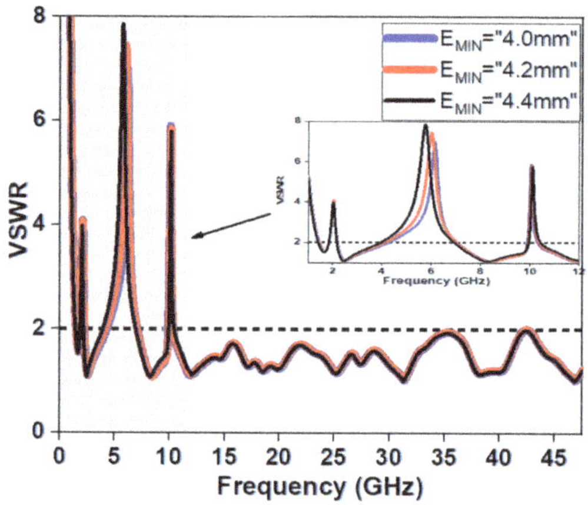

Figure 7. Parametric variation of independently controllable C band notch by varying E_{MIN}.

For a split elliptical slot of major axis length EMAX, minor axis length EMIN, and width ER, the design equations for calculating a band-notch at a given frequency can be represented as Equation (4).

$$E_c = D_e \pi (0.5E_{MIN} - E_R) = \frac{\lambda_g}{2} = \frac{c}{2f_{notch}\sqrt{\varepsilon_{eff}}} \tag{4}$$

$$D_e = 3(1+d) - \sqrt{(3+d)(1+3d)} \tag{5}$$

$$\varepsilon_{eff} = \frac{\varepsilon_r + 1}{2} + \frac{\varepsilon_r - 1}{2}\left[1 + 12\frac{h}{W_F}\right]^{-\frac{1}{2}} \tag{6}$$

where, E_c represents the inner perimeter of the elliptical slot, which could be equal to half of the directional wavelength at the intended notch frequency. The parameter D_e, utilized for the estimation of the perimeter of ellipse, is associated with the ellipticity by "d" through Equation (5) (where $d = E_{MAX}/E_{MIN}$ = 1.94), which is obtained as 3.01. The effective dielectric constant is calculated through Equation (6) as 1.84, where h (1.57 mm), W_F (3.6 mm), "ε_r" (2.2) are the substrate height, width of the microstrip feed, and relative permittivity of substrate, respectively. In the design simulations, the width of the slot was fixed at E_R = 0.2 mm. To obtain this second band-notch at 6 GHz within the C band (4–7.2 GHz), we chose E_{MIN} = 4.2 mm and E_{MAX} = 8.16 mm. The theoretically computed elliptical slot value (E_c) for 6 GHz was 18.43 mm, whereas the optimal parameter (E_c) for the slot was recorded as 18.02 mm.

3.3. Independently Controllable Notch X Band (9.8–10.4 GHz)

A C-shaped parasitic resonator was placed near the feed line for mitigating interference at X-band (9.8–10.4 GHz). Similar to the monitoring parameters of the AWS notch band and "C" notch band, parameter C_2 was the controlling parameter for the X band.

By altering the values of the parameter C_2, the third frequency notch band was shifted, whereas the first and second frequency notch bands remained unaltered. As is depicted in Figure 8, reducing the value of parameter C_2 displaced the third notch band towards higher frequencies. The third frequency notch band present at 10.38 GHz for C_2 = 2.2 mm was displaced to 10.22 GHz for C_2 = 2.3 mm and was shifted to 10 GHz for C_2 = 2.4 mm. From the parametric study it was found that C_2 had a considerable influence on localizing the third frequency notch band, whereas the remaining notch bands were unaltered. At the intended notched frequency, the length of the parasitic resonator is calculated using the expression given below in Equation (7).

$$L_{parasitic\ resonator} = \frac{\lambda_g}{2} = \frac{\lambda_0}{2\sqrt{\varepsilon_{reff}}} = \frac{c}{2f_{notch}\sqrt{\varepsilon_{reff}}} \quad (7)$$

$$\varepsilon_{eff} = \frac{\varepsilon_r + 1}{2} + \frac{\varepsilon_r - 1}{2}\left[1 + 12\frac{h}{W}\right]^{-\frac{1}{2}} \quad (8)$$

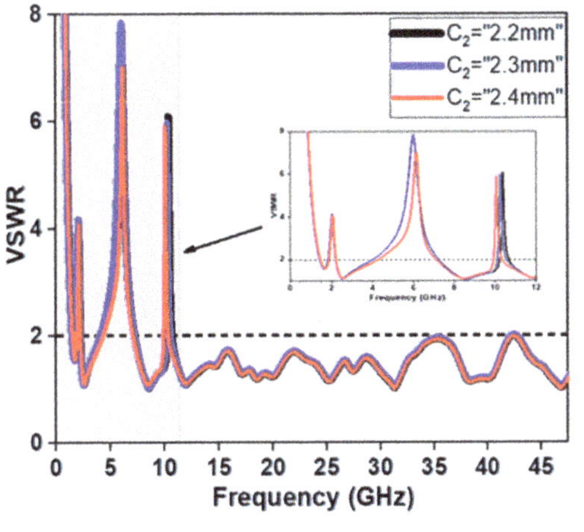

Figure 8. Parametric variation of the independently controllable C band notch by varying C_2.

The length of the parasitic resonator is equal to $C_1 + 2C_2 + 2C_3 - 4C_T$, where C_1 = 5 mm, C_2 = 2.4, C_3 = 1.6 mm, C_T = 0.54 mm, C_G = 0.6 mm, $\lambda_0 = c/f_{notch}$ is the free space wavelength, " c" is the speed of light, " f_{notch}" is the center frequency of the notched band, "ε_{eff}" is the effective permittivity, " ε_r" is the dielectric constant of the substrate, " h" is the height of the substrate, and W_F is the width of the feed line. The theoretically calculated C-shaped resonator value for 10 GHz is 11.05 mm, while the adjusted value for the resonator is 10.84 mm.

4. Current Distribution Analysis

To realize the creation of notch bands, the current density distribution at three notched frequency bands for AWS, C band, and X band was studied. Figure 9 illustrates the respective elements responsible for the band-notch characteristics. Strong current concentrations around slot, stub, and resonator inferred that there existed high mismatch in impedance, which indicated the creation of notch resonance at corresponding notch frequencies. Figure 9a shows that the maximum current was concentrated within the

E-shaped stub intended to notch the AWS band. Figure 9b depicts the surface currents at 6 GHz; the current was largely distributed around the elliptical-shaped slot compared to the other part of the antenna. Figure 9c gives the surface current distribution in the antenna at 10 GHz. A strong current dispersion was examined at the C-shaped resonator for the 10 GHz notched band, which further elucidated that this notch was produced due of the adverse effect of current caused by the C-shaped resonator.

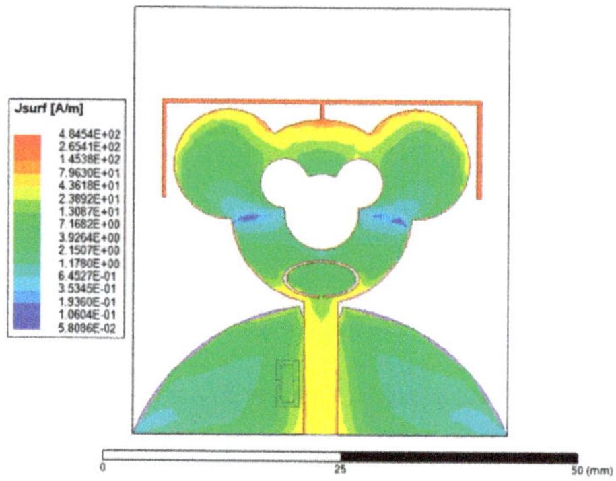

(**a**)

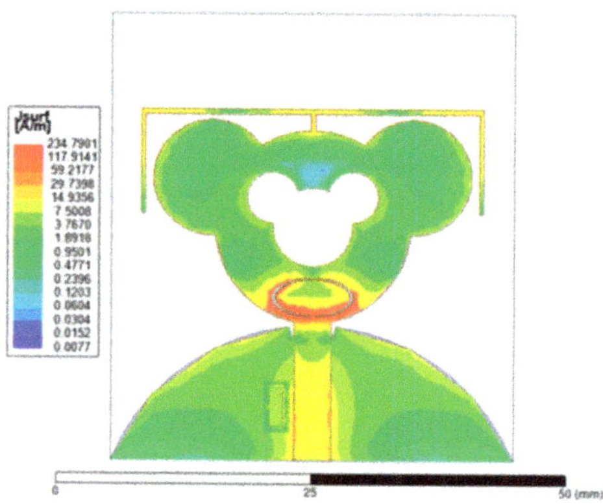

(**b**)

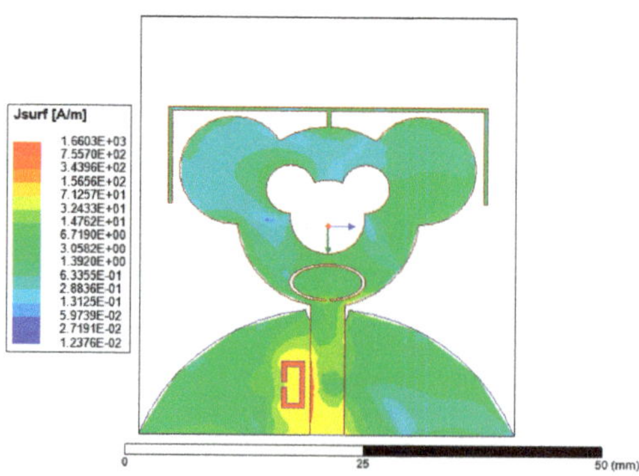

(**c**)

Figure 9. Simulated surface current distribution at the notch frequency (**a**) at 2 GHz with E-shaped stub, (**b**) at 6 GHz with split elliptical slot, and (**c**) at 10 GHz with C-shaped resonator.

5. Results and Discussion

In order to exhibit the efficacy and fulfillment of the designed antenna, a prototype was developed, and experimental measurements were carried out. Figure 10 shows the fabricated triple band-notched SWB antenna in reference to the above-mentioned parameters in Table 1. An Anritsu MS46122B vector network analyzer (VNA) was used to measure the antenna parameters. Even though the designed antenna functioned well for the frequency spectrum of 1.67–47.5 GHz, the empirical results were recorded only up to 20 GHz because of the higher frequency constraints of the available VNA.

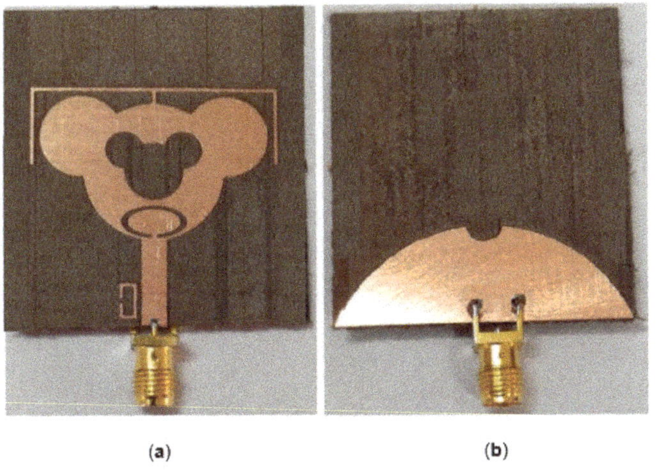

Figure 10. Fabricated antenna showing (**a**) E-shaped stub, split elliptical slot, and C-shaped resonator on Rogers's board (front view), and (**b**) elliptical ground plane (rear view).

Both simulated and measured VSWR results of the presented antenna are provided in Figure 11. Simulated and measured result showed that the designed antenna operated at 1.67–47.5 GHz with VSWR below 2, except the triple notched bands at 1.8–2.2 GHz, 4–7.2 GHz, and 9.8–10.4 GHz for rejecting the AWS and C and X band signals, respectively.

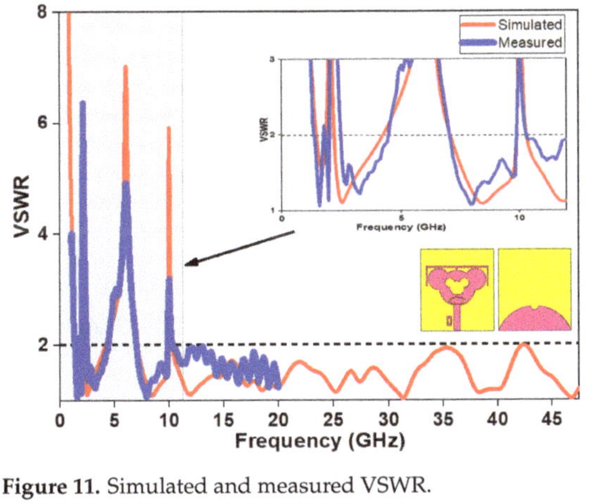

Figure 11. Simulated and measured VSWR.

5.1. Gain

The simulated 3D polar gain plot of the designed SWB antenna at the notch resonance frequencies of 2 GHz, 6 GHz, and 10 GHz is presented in Figure 12. Three sharp reductions in the gain at about −9.9 dB, −3.6 dB, and −1.8 dB in the notched bands at 2, 6, and 10 GHz, respectively, were perceived, which evidently proved the signal rejecting ability of the presented antenna.

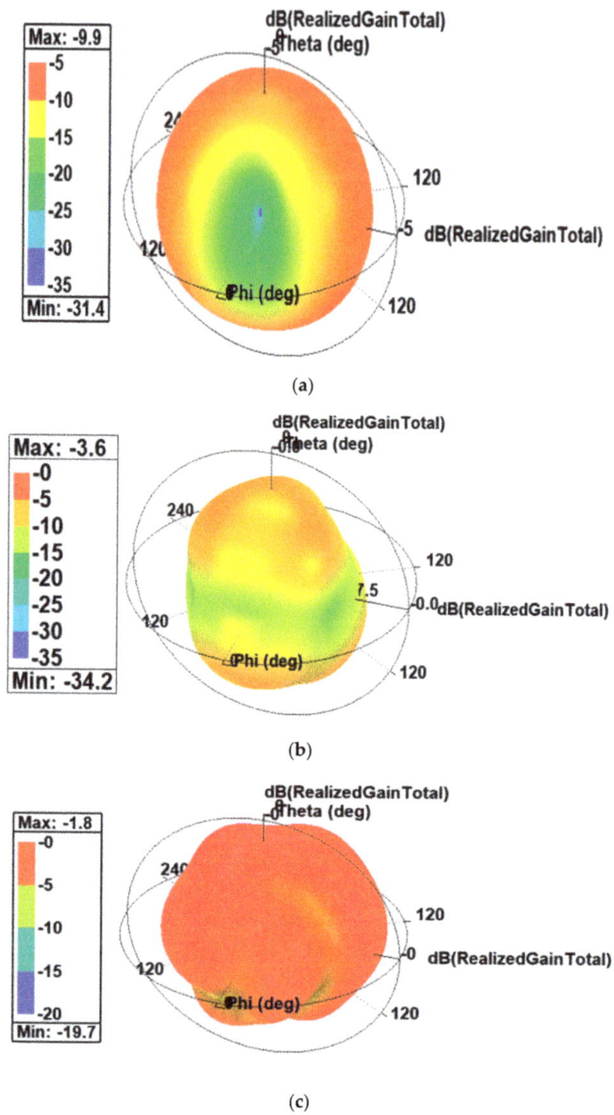

Figure 12. Three-dimensional polar gain plot for the designed antenna at notch frequency (**a**) 2 GHz, (**b**) 6 GHz, and (**c**) 10 GHz.

The gain obtained by simulating the model were proved to be in a positive agreement with measured results achieved from the fabricated prototype of the antenna. Figure 13 exhibits the variation of simulated and measured peak gain with frequency for the proposed antenna over the SWB frequency range (1.67–47.5 GHz). Abrupt gain reductions in the notched band could be anticipated because of elevated VSWR values at these frequencies, which resulted in low gain. At frequencies other than notched bands, the gain of the antenna was enhanced with frequency and increased from 1.7 dB to 10 dB. At notched frequencies, the simulated gain was −9.9 dB, −3.6 dB, and −1.8 dB, whereas the measured gain was −10.38 dB, −4.45 dB, and −2.08 dB at 2 GHz, 6 GHz, and 10 GHz, respectively. (Note: The empirical results were recorded only up to 20 GHz because of the higher frequency constraint of the available VNA).

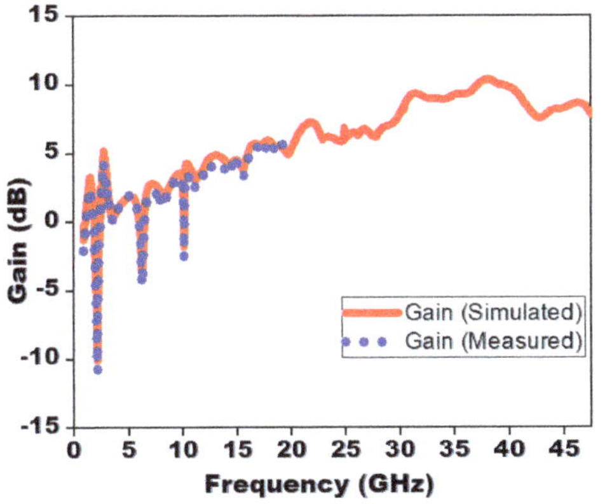

Figure 13. Simulated and measured peak gain of proposed antenna with notched bands.

5.2. Radiation Pattern

With a view to demonstrating that the designed antenna radiated across a broad frequency range, Figure 14 illustrates the simulated and experimented radiation patterns in the H-plane ($\varnothing = 90^{\circ}$) and E-plane ($\varnothing = 0^{\circ}$) for the passband resonance frequency of 2.53 GHz, 8.43 GHz, 11.88 GHz, and 18.51 GHz. It can be seen that the proposed antenna presented reasonably good bidirectional monopole E-plane patterns and omni-directional H-plane patterns over the different passband frequencies. At the higher frequencies (greater than 8 GHz), the radiation patterns in both planes were distorted on account of the higher order modes' excitation and non-uniform phase division. As a consequence, radiated modes may have promoted an increased level of electromagnetic interference (EM). In order to suppress these unwanted radiated modes, a designer can employ techniques such as introducing periodic structures, parasitic elements, split ring resonators (SRRs), etc. However, these techniques increase the complexity of the design and cost of fabrication.

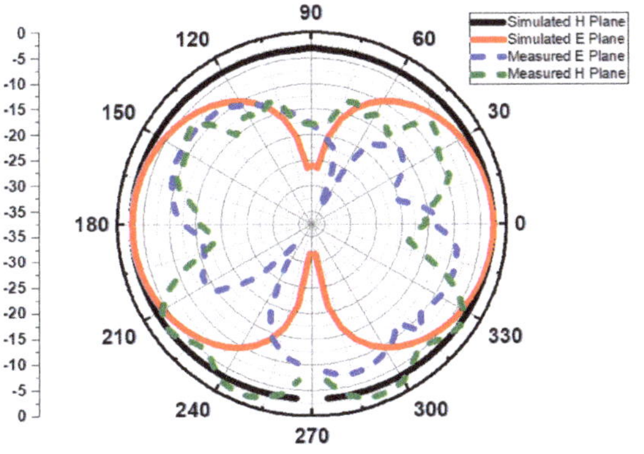

(a)

Figure 14. *Cont.*

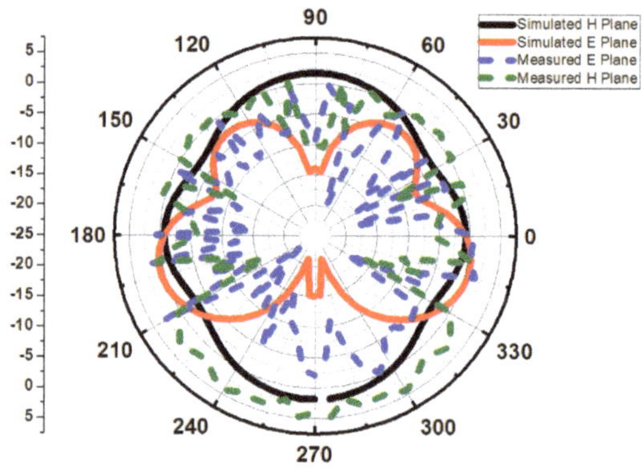

(b)

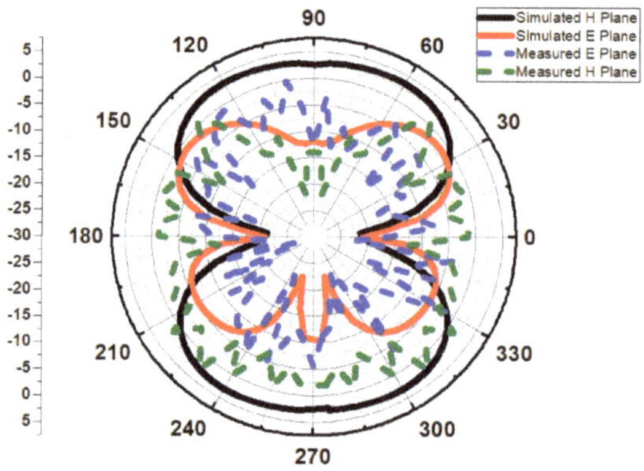

(c)

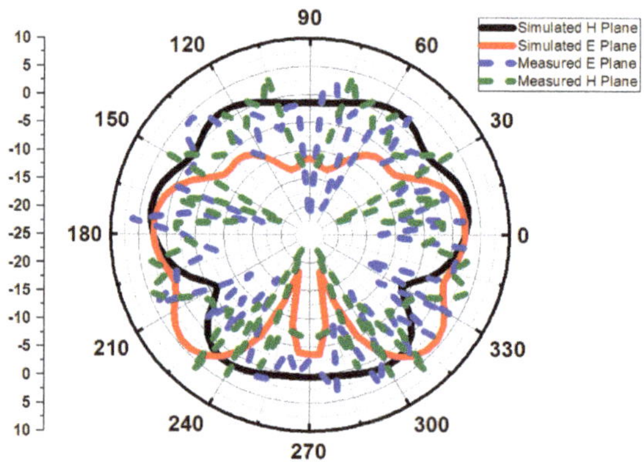

(d)

Figure 14. Simulated and measured far field radiation patterns of presented antenna at (∅ = 90°) and (∅ = 0°) for (**a**) 2.53 GHz, (**b**) 8.43 GHz, (**c**) 11.88 GHz, and (**d**) 18.51 GHz.

6. Time-Domain Characterization

SWB antennas must be capable of operating over the entire bandwidth with good efficiency and acceptable radiation properties. Simultaneously, the transmitted pulse from the SWB antenna should have a good impulse response with minimum distortion. In addition, it is necessary that the pulse be identified at the receiver end with minimal distortion. Therefore, to endorse the significance of the antenna for pulsed communication systems, time-domain analysis was performed. The time-domain characterization of the antenna parameters confirmed that the designed antenna was an excellent candidate for triple band-notched SWB applications. To realize the time-domain behavior of the signal at band-notched frequencies, two similar antennas were arranged in face-to-face and side-by-side configurations 75 mm apart, as shown in Figure 15. In both the configurations, Gaussian pulses, as shown in Figure 16, were taken into consideration. A few key time-domain metrics were investigated in this following subsection.

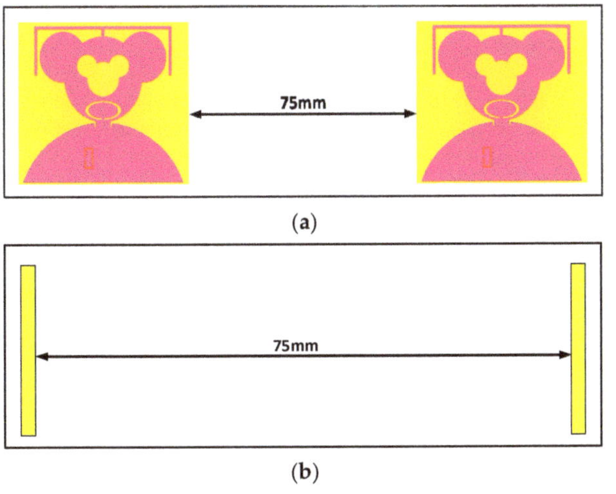

Figure 15. Two different orientations of the presented radiator for time-domain analysis: (**a**) side-by-side and (**b**) face-to-face orientations.

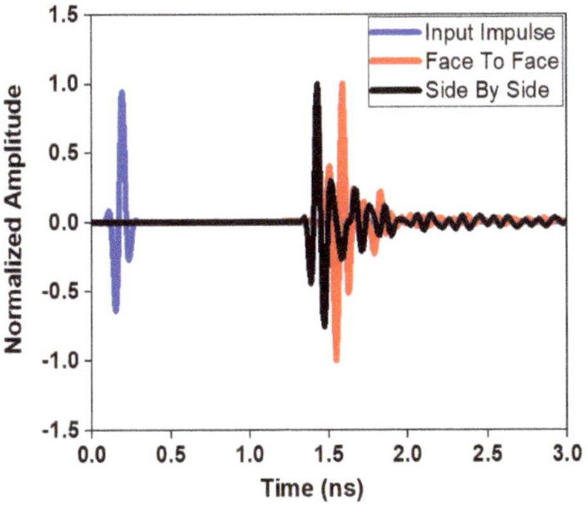

Figure 16. Normalized received pulse for triple band-notched SWB antenna system in face-to-face and side-by-side arrangements.

6.1. System Fidelity Factor (SFF)

SSF can be described as the cross-correlation between the normalized input waveform $T_s(t)$ and normalized received waveform $R_s(t)$ and can be evaluated by Equation (9)

$$SFF = \max_{T} \int_{-\infty}^{\infty} \hat{T}_s(t)\hat{R}_s(t+\tau)dt \quad (9)$$

where

$$\hat{T}_s(t) = \frac{T_s(t)}{\left[\int_{-\infty}^{\infty} |T_s(t)|^2 dt\right]^{\frac{1}{2}}} \tag{10}$$

$$\hat{R}_s(t) = \frac{R_s(t)}{\left[\int_{-\infty}^{\infty} |R_s(t)|^2 dt\right]^{\frac{1}{2}}} \tag{11}$$

where, $T_s(t)$ is the normalized transmitted signal pulse, and $R_s(t)$ is the normalized received signal pulse for the presented triple band-notched SWB antenna.

Normalization is performed to ensure that only the shape of the signals is considered without taking its magnitude into account. The ideal value of SFF lies between 0 and 1. If it is 0, then the received signal is completely different from transmitted signal, and if it is 1, then both the transmitted and received signals are identical. The signal is unrecognizable when the $SFF < 0.5$. Hence, a practical SFF should be near to 1 for ensuring reliable antenna performance over wide frequency ranges.

The SFF is evaluated by finding the similarity between $T_s(t)$ and $R_s(t)$ using Equation (9). The calculated SFFs for the triple band-notched SWB antenna system for face-to-face and side-by-side configurations were 93% and 90%, respectively.

As depicted in Figure 16, the undesired oscillations in the received signal for the triple band-notched SWB antenna system in both configurations were considerably small. It could be observed that the received signal had lower distortion and relatively maintained the shape of the transmitted pulse. This confirmed that the received signal was less prone to inter symbol interference (ISI).

6.2. Group Delay

The amount of distortion in the SWB antenna can be defined in terms of the group delay of the device. A uniform sustained group delay is always desirable for the entire SWB frequency range. The group delay of the proposed antenna system was steady, except at the notch bands. There was a sudden variation in the group delay at the notch frequency for the band-notched SWB antenna system due to the appearance of phase irregularities. From Figure 17, it can be seen that a high group delay was observed at the notched frequencies (2, 6, and 10 GHz), whereas for passband frequencies, group delays for the face-to-face and side to side arrangements were within acceptable limit of less than 1 ns.

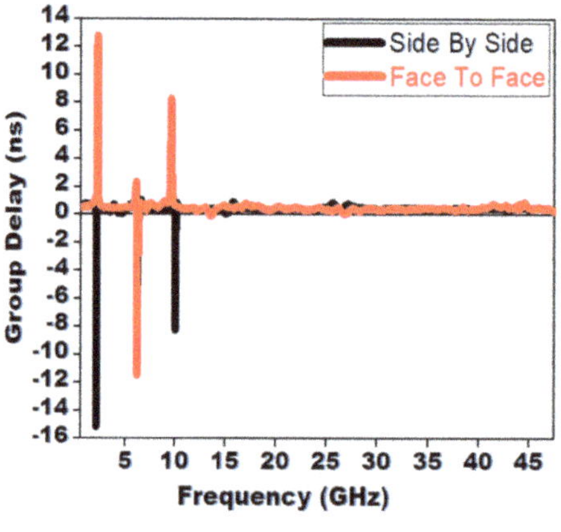

Figure 17. Group delay for proposed band notch antenna for side-by-side and face-to-face configurations.

6.3. Transfer Function

Since the SWB systems use short pulses to transmit and receive signals, it was essential to study the transfer functions for evaluating the proposed antenna's performance and designing the transmitted pulse signals.

In the transfer function (S21) of the band-notched SWB antenna system, there was a sharp dip at the notch frequencies. From Figure 18, it was observed that the magnitude of S21 in both configurations was flat over the entire SWB spectrum, except at band-notched regions. In the case of the face-to-face orientation, attenuations of 54 dB, 60 dB, and 37 dB were observed at the first notched band centered at 2 GHz, the second notched band at the 6 GHz region, and the third notched band at the 10 GHz region, respectively. In the side-by-side configuration, the attenuation was comparably less than the face-to-face configuration at band-notched regions.

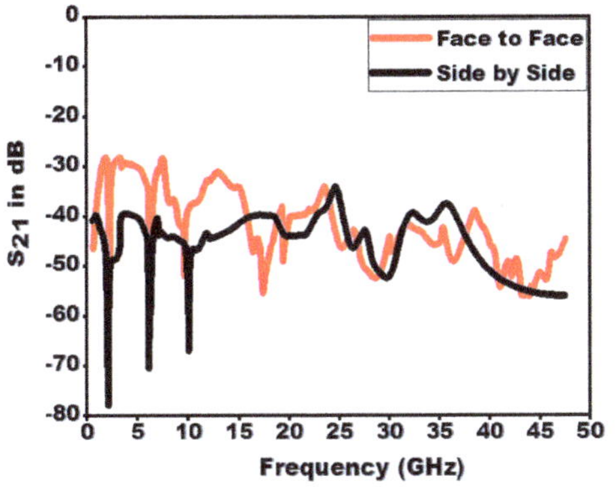

Figure 18. Transfer function (S_{21}) for the proposed triple band notch antenna for face-to-face and side-by-side configuration.

7. Comparative Analysis

Some distinct techniques were already well demonstrated in achieving a wideband antenna with single/dual/triple band-notch characteristic [1–15,17–25] in the introductory section of this article. Thus, it is worth comparing the proposed design with the existing designs stated in the literature as presented in Table 3. It can also be seen from Table 3 that the presented antenna has considerably larger bandwidth ratios than the other referenced antennas. Although the size of the antenna is somewhat larger than those in the reported literature, this is nevertheless the best one with respect to size of frequency range compared to other works. The main advantage of the proposed antenna is that it can be tuned independently, ensuring that the notch frequencies are minimally mutually coupled. Rejection levels of −9.9 dB, −3.6 dB, and −1.8 dB were encountered at notch frequencies of 2 GHz, 6 GHz, and 10 GHz, respectively, which confirms the efficacy of band-notched features of this presented design. The higher the rejection levels, the better the triple band notch antenna performance is. It could be observed that the proposed antenna showed superior performance compared to most of the previous studies in terms of bandwidth, gain at passband, ratio bandwidth, bandwidth percentage, and bandwidth dimension ratio. Therefore, the proposed triple band-notched design is a prominent radiating structure for SWB applications.

Table 3. Comparison of proposed design with existing designs.

Ref. No.	Size (λ Free Space Wavelength Calculated at Lower Frequency)	Frequency (GHz)	Notch BW (GHz)	Notch Bands	BW (%)	Bandwidth Ratio	BDR	Gain (Passband in dB)	Gain at Notch Frequency (Rejection Level in dB)	VSWR at the Notch Band	No. of Notches	Antenna Structure	
												Radiator Design	Notch Structure
[1]	0.11λ × 0.15λ	2.8–13	3.3–4.2, 5.1–5.4	WiMAX, WLAN,	129	4.64:1	7818	0–2.6	5, 6	7,5	Dual	Square-shaped radiator	J-shaped and F-shaped slot
[2]	0.07λ × 0.11λ	2.12–14.8	3.36–4.16, 4.92–5.36, 5.68–6.0	WiMAX, lower WLAN, upper WLAN	150	6.98:1	19,480	2–6	12, 9, 8	10,5,4	Triple	Rectangular patch antenna	Inverted T-shaped slot, comb-shaped slot, rectangular-shaped slot
[3]	0.24λ × 0.29λ	2.8–12.6	3.43–3.65, 4.95–5.25, 5.36–5.85	WiMAX, WLAN	127	4.5:1	1825	2.5–5.5	2.5, 2, 0.5	5,5.5,3.5	Triple	Circular patch radiator	C-shaped slots
[4]	0.35λ × 0.35λ	3–11	5.15– 5.35, 5.725–5.825	WLAN	114	3.66:1	931	2.5–4	(−5.2 and −2.4)	5.8,8.5	Dual	Circular radiating patch	Quarter wavelength open ended slits
[5]	0.30λ × 0.28λ	3–11	5.0–5.8	WLAN	114	3.66:1	1357	2.5–7	12.5	27	Single	Circular-shaped radiating patch	Pair of half wavelength stubs and slits
[6]	0.3λ × 0.24λ	2.8–11.0	3.3–4, 5.05–5.90	WiMAX, WLAN	119	3.92:1	1653	1–4	4, 6	15,11	Dual	Square radiating patch	T-shaped stub, pair of U-shaped parasitic strips
[7]	0.24λ × 0.24λ	2.39–2.49, 3.1–11.4	4.97–5.48, 5.69–5.99	Lower WLAN, Upper WLAN	130	4.76:1	2257	0.5–5	4, 1	Not defined	Dual	Quasi U-shaped antenna	Pair of inverted L-shaped stubs
[8]	0.25λ × 0.29λ	2.76–39.53	3.74, 5.45, 7.65	WiMAX, WLAN X band	174	14.32:1	2400	5–6.88	11.38, 9.65, 8.13	20.64, 5.65 and 12.05	Triple	Guitar-shaped radiating patch	Inverted T-shaped stub, pair of elliptical slots
[9]	0.15λ × 0.22λ	2.34–20.00	3.30–3.80, 5.150–5.825, 7.25–7.75	WiMAX, WLAN, DSS	158	8.54:1	4788	3.39–4.98	6.95, 8.95, 7.98	Not defined	Triple	Tilted ellipse-shaped radiating patch	Fractal stub, pair of modified rectangular slots
[10]	0.26λ × 0.35λ	2.68–11.15	3.3–3.6, 5.15–5.35, 5.725–5.825	WiMAX, WLAN, Upper WLAN	122	4.16:1	1341	2–4.75	3.4, 2.3 and 2.1	Not defined	Triple	Elliptical-shaped radiating patch	Three sets of meandered ground stubs
[11]	0.32λ × 0.24λ	3–11	5–5.4, 7.8–8.4	WLAN ITU band	114	3.66:1	1484	1.79–4.47	2, 3	Not defined	Dual	Circular monopole radiator	U-shaped and I-shaped parasitic elements
[12]	0.13λ × 0.27λ	1.6–25	3.2–4.4, 7.2–8.4	WiMAX/C band	176	15.6:1	5014	5.8	14, 8	5,4	Dual	Circular monopole radiating patch	U-shaped parasitic element, U-shaped slot
[13]	0.2λ × 0.2λ	3–17	5.15–5.825	IEEE802.11a, HIPERLAN/2	140	5.66:1	3500	2–5	4	4	Single	Hemi-ellipse-shaped antenna	Parasitic strip
[14]	0.26λ × 0.26λ	2.21–11.71	3.3–3.8, 5.15–5.85, 7.9–8.4	WiMAX, WLAN, X-band	136	5.29:1	2011	1–6	1, 3	7.5, 5.5, 4	Triple	Semi-elliptical-shaped radiating patch	Complementary SRR and rectangular SRR
[15]	0.24λ × 0.35	3.1–11	3.4–3.6, 5.1–5.3, 5.7–5.9	WiMAX WLAN	112	3.54:1	1333	2–4.5	6, 2.5, 2	Not defined	Triple	Bevel-shaped radiating patch	Pair of slots, circular CSRR, rectangular-shaped CSRR
[17]	0.17λ × 0.17λ	2.8–14.8	3.3–3.6, 5.15–5.825	WiMAX WLAN	136	5.28:1	4706	0.5–5	4, 6	18, 11	Dual	Square-shaped radiating patch	Inverted L-shaped slot, U-shaped slot
[18]	0.32λ × 0.29λ	2.4–21.7	5.15–5.825, 3.3–3.6	IEEE802.11a, HIPERLAN/2 WLAN, WiMAX	153	9.04:1	1649	Not defined	Not defined	4.5, 4.5	Dual	Stepped square-shaped radiating patch	L-shaped slot, U-shaped slot
[19]	0.21λ × 0.26λ	2.45–12	3.27–3.57, 5.01–5.45, 5.55–6.05, 7.05–7.45, 7.83–8.19	WiMAX, WLAN, downlink of X-band satellite communication, ITU8-GHz band signals	132	4.89:1	2418	2–6	2, 3.0.5, 1, 1.5	5, 7, 8, 5, 8	Penta	Portrait radiating patch	C-shaped slot, nested C-shaped slot, arc-shaped stub, pair of U-shaped slots
[20]	0.15λ × 0.16λ	1.4–11.3	1.8–2.3, 5.6–6.1, 3.2–3.8	Advanced wireless system, IEEE 802.11/HIPERLAN, WiMAX	156	8.07:1	6500	3.6–4.6	2.5, 3, 1	9, 10, 18	Triple	Maple leaf-shaped radiator	E-shaped resonator, meandered slot, and U-shaped slot
[21]	0.35λ × 0.33λ	2.9–13	3.3–3.9, 5.2–5.35, 5.8–6.0	WiMAX WLAN	127%	4.48:1	1099	4–6	5, 6, 7	20, 10, 20	Triple	Spade-shaped monopole radiator	Hook-shaped structure, inverted ohm-shaped slot, semi octagonal resonator
[22]	0.43 × 0.61	2.6–10.8	5.09, 6.34, and 8.04	WLAN X Band	122	4.15:1	465	0–4	9.6, 12.7,4.27	Not defined	Triple	Circular monopole radiating patch	Three pairs of circular SRRs,
[23]	0.33λ × 0.29λ	2.58–11.62	3.28–3.82, 5.12–5.4, 5.7–6	WiMAX, WLAN, DSRC	127	4.5:1	1327	2–5	5, 10, 4	12, 10, 4	Triple	Leaf structured CPW-fed ground plane	Two rectangular SRRs and one circular SRR
[24]	0.43λ × 0.24λ	3.1–11	3.3–3.7, 7.1–7.76, 5.15–5.825	WiMAX, Downlink X-band satellite communication, WLAN	112	3.54:1	1085	2.7–3.6	5, 2, 6	Not defined	Triple	Fork-shaped radiating patch	U-shaped slot, modified U-shaped slot, semi-arch-shaped slot
[25]	0.35λ × 0.36λ	0.72–25	2.4–2.484, 5.15–5.85, 7.25–7.75	WLAN, downlinks of X-band satellite communication	189	34.72:1	1500	2.5–7.5	1.5, −2, 3	10, 10, 10	Triple	Modified elliptical-shaped radiating patch	Pair of arc-shaped slots, inverted U-shaped slot
Prop. work	0.22λ × 0.25λ	1.67–47.5	1.8–2.2, 4–7.2, 9.8–10.4	AWS, C Band, X Band	186	28.44:1	3382	2–15	9.9, 3.6, 1.8	4, 7, 5.9	Triple	Concentric structured monopole antenna	E-shaped stub, split elliptical slot, C-shaped resonator

8. Conclusions

A compact concentric structured SWB antenna with triple notch bands using an E-shaped stub, split elliptical-shaped slot, and C-shaped resonator has been demonstrated experimentally. The presented antenna offers a broad impedance bandwidth of 45.83 GHz (1.67–47.5 GHz), with triple notches at 1.8–2.2 GHz, 4–7.2 GHz, and 9.8–10.4 GHz to eliminate the interference from co-existing AWS, C, and X bands, respectively. Based on empirical formulas, three notch bands are developed at the intended frequencies with appropriate dimensions. In order to understand the triple band-notched functions, the surface current distribution at the band-notched frequencies are computed and examined. The band notches of the designed antenna can be tuned autonomously through their particular governing parameters. The simulated and measured results of VSWR, gain, and radiation pattern are compared, and a close agreement between simulated and measured results is observed. The presented antenna has an elementary linear structure with a concise dimension of 45×40 mm^2, allowing it to be embedded within portable devices conveniently. The designed antenna offers excellent time-domain performances, such as constant group delay, minimum pulse distortion, and an abruptly reduced gain at the stop bands, which makes it appropriate for pulse-based SWB communications.

Author Contributions: Conceptualization: T.A. and W.B.; methodology: T.A., W.B., A.S., P.K., S.D. and J.A.; software: W.B. and A.S.; validation, M.S., W.B., T.A., P.K., A.S., S.D. and J.A.; data citation: W.B. and A.S.; writing: W.B., T.A., S.D. and A.S.; supervision: T.A., M.S., P.K., and J.A. All authors have read and agreed to the published version of the manuscript.

Funding: This research received no external funding.

Institutional Review Board Statement: Not applicable.

Informed Consent Statement: Not applicable.

Data Availability Statement: Not applicable.

Conflicts of Interest: The authors declare no conflict of interest.

References

1. Sahoo, S.; Mohanty, M.N.; Mishra, L.P. Bandwidth improvement of compact planar antenna for UWB application with dual notch band performance using parasitic resonant structure. *Prog. Electromagn. Res.* **2018**, *66*, 29–39.
2. Dong, J.; Li, Q.; Deng, L. Compact planar ultrawideband antennas with 3.5/5.2/5.8 GHz triple band-notched characteristics for internet of things applications. *Sensors* **2017**, *17*, 349. [CrossRef] [PubMed]
3. Xu, J.; Wang, G. A compact printed UWB antenna with triple band-notched characteristics. *Microw. Opt. Technol. Lett.* **2012**, *54*, 2146–2150. [CrossRef]
4. Zhou, D.; Gao, S.; Zhu, F.; Abd-Alhameed, R.A.; Xu, J.D. A simple and compact planar ultra wideband antenna with single or dual band-notched characteristics. *Prog. Electromagn. Res.* **2012**, *123*, 47–65. [CrossRef]
5. Shi, M.; Cui, L.; Liu, H.; Lv, M.; Sun, X.B. A new UWB antenna with band-notched characteristic. *Prog. Electromagn. Res.* **2018**, *74*, 201–209. [CrossRef]
6. Jiang, W.; Che, W. A novel UWB antenna with dual notched bands for WiMAX and WLAN applications. *IEEE Antennas Wirel. Propag. Lett.* **2012**, *11*, 293–296. [CrossRef]
7. Yang, B.; Qu, S. A compact integrated Bluetooth UWB dual-band notch antenna for automotive communications. *AEU Int. J. Electron. Commun.* **2017**, *80*, 104–113. [CrossRef]
8. Sharma, M. Superwideband Triple Notch Monopole Antenna for Multiple Wireless Applications. *Wirel. Pers. Commun.* **2019**, *104*, 459–470. [CrossRef]
9. Dhasarathan, V.; Sharma, M.; Kapil, M.; Vashist, P.C.; Patel, S.K.; Nguyen, T.K. Integrated bluetooth/LTE2600 superwideband monopole antenna with triple notched (WiMAX/WLAN/DSS) band characteristics for UWB/X/Ku band wireless network applications. *Wirel. Netw.* **2020**, *26*, 2845–2855. [CrossRef]
10. Weng, Y.F.; Cheung, S.W.; Yuk, T.I. Triple band-notched UWB antenna using meandered ground stubs. In Proceedings of the 2010 Loughborough Antennas & Propagation Conference 2010, Loughborough, UK, 8–9 November 2010; pp. 341–344.
11. Yadav, D.; Abegaonkar, M.P.; Koul, S.K.; Tiwari, V.; Bhatnagar, D. A compact dual band-notched UWB circular monopole antenna with parasitic resonators. *AEU Int. J. Electron. Commun.* **2018**, *84*, 313–320. [CrossRef]
12. Manohar, M.; Kshetrimayum, R.S.; Gogoi, A.K. A compact dual band-notched circular ring printed monopole antenna for super wideband applications. *Radio Eng.* **2017**, *26*, 64–70. [CrossRef]

13. Kim, K.H.; Park, S.O. Analysis of the small band-rejected antenna with the parasitic strip for UWB. *IEEE Trans. Antennas Propag.* **2006**, *54*, 1688–1692. [CrossRef]
14. Sarkar, D.; Srivastava, K.V.; Saurav, K. A compact microstrip-fed triple band-notched UWB monopole antenna. *IEEE Antennas Wirel. Propag. Lett.* **2014**, *13*, 396–399. [CrossRef]
15. Sarkar, M.; Dwari, S.; Daniel, A. Printed monopole antenna for ultra-wideband application with tunable triple band-notched characteristics. *Wirel. Pers. Commun.* **2015**, *84*, 2943–2954. [CrossRef]
16. Balani, W.; Sarvagya, M.; Samasgikar, A.; Ali, T.; Kumar, P. Design and Analysis of Super-Wideband Antenna for Microwave Applications. *Sensors* **2021**, *21*, 477. [CrossRef]
17. Esmati, Z.; Moosazadeh, M. Design of compact dual band-notched UWB antenna using slotted conductor-backed plane. *Arab. J. Sci. Eng.* **2014**, *39*, 4707–4713. [CrossRef]
18. Tilanthe, P.; Sharma, P.C.; Bandopadhyay, T.K. A compact UWB antenna with dual band rejection. *Progress Electromagn. Res.* **2011**, *35*, 389–405. [CrossRef]
19. Xu, J.; Shen, D.; Zhang, X.; Wu, K. A compact disc ultrawideband (UWB) antenna with quintuple band rejections. *IEEE Antennas Wirel. Propag. Lett.* **2012**, *11*, 1517–1520.
20. Iqbal, A.; Smida, A.; Mallat, N.K.; Islam, M.T.; Kim, S. A compact UWB antenna with independently controllable notch bands. *Sensors* **2019**, *19*, 1411. [CrossRef]
21. Li, W.T.; Shi, X.W.; Hei, Y.Q. Novel planar UWB monopole antenna with triple band-notched characteristics. *IEEE Antennas Wirel. Propag. Lett.* **2009**, *8*, 1094–1098.
22. Shaik, L.A.; Saha, C.; Siddiqui, J.Y.; Antar, Y.M. Ultra-wideband monopole antenna for multiband and wideband frequency notch and narrowband applications. *IET Microw. Antennas Propag.* **2016**, *10*, 1204–1211. [CrossRef]
23. Kundu, S.; Jana, S.K. Leaf-shaped CPW-fed UWB antenna with triple notch bands for ground penetrating radar applications. *Microw. Opt. Technol. Lett.* **2018**, *60*, 930–936. [CrossRef]
24. Elhabchi, M.; Srifi, M.N.; Touahni, R. A tri-band-notched UWB planar monopole antenna using DGS and semi arc-shaped slot for WIMAX/WLAN/X-band rejection. *Prog. Electromagn. Res.* **2017**, *70*, 7–14. [CrossRef]
25. Liu, J.; Esselle, K.P.; Hay, S.G.; Zhong, S.S. Study of an extremely wideband monopole antenna with triple band-notched characteristics. *Prog. Electromagn. Res.* **2012**, *123*, 143–158. [CrossRef]

applied sciences

MDPI

Article

Optimization of a Coherent Dual-Beam Array Feed Network for Aperiodic Concentric Ring Antennas

Armando Arce [1,2,*], Enrique Stevens-Navarro [2], Ulises Pineda-Rico [2], Marco Cardenas-Juarez [2], Francisco R. Castillo-Soria [2] and David H. Covarrubias [3]

1 Cátedras CONACYT, Universidad Autónoma de San Luis Potosí (UASLP), San Luis Potosí 78290, Mexico
2 Facultad de Ciencias, Universidad Autónoma de San Luis Potosí (UASLP), San Luis Potosí 78290, Mexico; estevens@galia.fc.uaslp.mx (E.S.-N.); u_pineda@fc.uaslp.mx (U.P.-R.); mcardenas@fc.uaslp.mx (M.C.-J.); ruben.soria@uaslp.mx (F.R.C.-S.)
3 Electronics and Telecommunications Department, CICESE Research Center, Carretera Ensenada-Tijuana No.3918, Zona Playitas, Ensenada 22860, Mexico; dacoro@cicese.mx
* Correspondence: armando.arce@uaslp.mx

Abstract: A dual-beam coherent feeding system design approach with a non-uniform layout on a concentric ring array is described and synthesized. In this case, the feeding system is based on a reconfigurable topology composed of a set of alternated power dividers and combiners, providing coherent in-phase outputs. Thus, in this paper, a two-beam architecture based on a coherent feeding system formed by a set of intercalated input signals feeding each circular ring in a non-uniform antenna array with multi-beam shaping and steering features is analyzed. The task of optimizing the aperiodic layout on the shared aperture based on the radii of the circular rings is realized by the differential evolution method. Numerical experiments grounded in antenna synthesis validate the capabilities and improved performance of the proposed dual-beam configuration with a non-uniform layout in contrast with its uniform counterpart, with enhanced performance on average by up to −6.1 dB for sidelobe level and 3.5 dB for directivity. Additionally, the results show a significantly less complex two-beam feeding network in contrast with the case of a typical electronically scanned array—in this proposal, each direction of maximum radiation is conformed and scanned with approximately half of the control inputs.

Keywords: beamforming networks; concentric rings geometry; aperiodic arrays; dual-beam; antenna synthesis; optimization; evolutionary computing

Citation: Arce, A.; Stevens-Navarro, E.; Pineda-Rico, U.; Cardenas-Juarez, M.; Castillo-Soria, F.R.; Covarrubias, D.H. Optimization of a Coherent Dual-Beam Array Feed Network for Aperiodic Concentric Ring Antennas. *Appl. Sci.* **2021**, *11*, 1111. https://doi.org/10.3390/app11031111

Academic Editor: Jaume Anguera
Received: 3 December 2020
Accepted: 21 January 2021
Published: 26 January 2021

1. Introduction

Arrays of antennas are a fundamental actor and base in wireless communication systems, where a group of radiators can perform advanced features synthesizing one or more beam patterns of arbitrary shape. These antenna arrays extend the capabilities of mono-element antennas, improving their overall performance. Hence, increasing the number of antennas can improve the antenna system's capabilities in some cases with multiple shaped and directive beams, power pooling, element redundancy, angular diversity, and electronic beam scanning (phased arrays). However, these multi-beam antennas are not a single entity and their function relies on an underlying feeding system, usually a beamforming network (BFN) subsystem that is an essential component for feeding and combining the array element's signals [1]. For fixed beams, a single high directive antenna without a feeding network could also be an alternative to arrays [2].

As a result of the continuous advancement of telecommunication systems (e.g., 5G mobile communications and beyond), a real need and motivation have arisen focused on the evolution and improvement of antenna systems. In light of this, the main goal of research in this field has been to increase the global antenna performance and at the same time reduce the major issues in these systems related to cost and complexity [3,4]. Within this framework, the aperiodicity of antenna arrays could be a solution to enhance the antenna system's capabilities and even minimize the use of some radiators. Nevertheless, the design and optimization of aperiodic arrays usually become a complex target, where in addition to the complexity of the aperiodicity of the layout, the inclusion of a multiple-beam feeding system in the synthesis makes this task more challenging.

Specifically, dual-beam antenna systems are useful in different mobile applications including terrestrial and space systems, such as microwave close-range sensors, automotive radar sensors, synthetic aperture radar (SAR) systems, tracking satellites, autonomous sensor networks, and base stations, among others [5,6]. Nevertheless, there are varying approaches to conform two beams with pros and cons, and trade-offs between them. Some antenna system architectures involve the use of certain types of radiators and feeders working together [7]—typical examples include planar leaky wave antennas (LWA) with different methods of feeding network to control the leaking rate, direction, and polarization [8]. Similarly, transmit arrays as a beamforming network are another example of a dual-beam solution grounded on the behavior of frequency selective surfaces and transmitting/receiving elements [9,10]. In the same vein, phased arrays are a cornerstone for dual/multi-beam beamforming and beam steering in infrastructure applications [11], sometimes combined with reflectarray apertures [12] or even feeding the array with a peripherally excited (PEX) metallic cavity [4]. Some other typical applications use classical lossless matrices or hybrids using related circuitry as beamforming networks [13,14].

However, the aforementioned dual-beam approaches deal with different drawbacks, including a cumbersome mechanism for beam scanning or fixed beams, interference between signals caused by adjacent antennas, and limited beam pointing or beam shaping, among others. Nevertheless, an alternative proposal is based on coherent in-phase networks, improving flexibility and reconfigurability with refined beam shaping and scanning (electronically) capabilities. In previous years, a coherent feeding system relying on identical substructure components known as coherently radiating periodic structures was presented in [15]. The feeding system for a phased array was described in [16], showing the working principles to implement a beamforming network topology in a basic antenna array, minimizing the difficulty of the associated control to scan the beam. As an antenna synthesis problem with optimization, the antenna system with a periodic aperture in different antenna geometries and parameters has been widely studied, and some related investigations can be found in [17–19]. Recently, early work on non-uniform apertures working together with coherent feeding systems in circular antenna arrays was addressed in [20] and a mono-beam investigation in concentric rings was analyzed in [21].

This paper presents a dual-beam analysis of an antenna system's performance that exploits a non-uniform aperture with concentric ring antennas, employing a minimalistic architecture based on coherently radiating periodic structures with multi-beam steering capabilities. In this approach, the antenna system is optimized in such a way that the radii on the antenna array are improved by a well-known evolutionary optimizer. Furthermore, a benchmarking study with a periodic aperture and the same feeding system with contrasting directivity (D) and sidelobe level (SLL) is implemented and validated by the antenna synthesis' numerical experiments.

The remainder of this paper is structured as follows: the mathematical expressions of a non-uniform layout on concentric ring antennas, the general theory behind coherent periodic structures acting as a beamforming network, and the methodology to optimize the antenna system are presented in Section 2. Section 3 addresses the numerical examples set out for the two-beam feeding system configuration and the parameters used in the metaheuristic optimizer, and at the end, the numerical examples for the dual-beam array feed are assessed and reviewed. Some concluding remarks are presented in Section 4.

2. Problem Formulation

2.1. Aperiodicity in Concentric Ring Arrays

An aperiodic or non-uniform array is considered when the interelement spacing of antenna elements is not equal between them. Thus, consecutive rings conformed by antenna elements in a circular layout which share the same origin where each ring has its own radius within an aperiodic layout (different interelement distances or radii between rings) are known as aperiodic concentric ring arrays (ACRAs). In general, let us consider a concentric geometry, assuming N_r circular rings with N_e radiators for each ring which are deployed in the $x - y$ plane, as illustrated in Figure 1. The far-field pattern considering an arbitrary complex excitation set (w_{mn}) without a radiator in the center of the array can be calculated as [22]:

$$AF(\theta,\varphi) = \sum_{n=1}^{N_r}\sum_{m=1}^{N_e} w_{mn}\exp[jk[x_{nm}(u) + y_{nm}(v)]], \quad (1)$$

where $x_{nm} = r_n \cos\varphi_{nm}$ and $y_{nm} = r_n \sin\varphi_{nm}$ define the antenna element locations, the antenna elements' deployment around the rings is indicated by $\varphi_{nm} = 2\pi(m-1)/N_e$, and r_n denotes the radial distance per ring from the origin of the antenna array until the last ring. The $u - v$ space considering the excitation phases is defined by $u = \sin\theta\cos\varphi - \sin\theta_0\cos\varphi_0$ and $v = \sin\theta\sin\varphi - \sin\theta_0\sin\varphi_0$. The steering angle for translation is denoted by (θ_0, φ_0); in this case, θ is the angular direction in elevation and φ is the spatial angle for the azimuth. Finally, the wavelength λ is related to the angular wavenumber $k = 2\pi/\lambda$ in Equation (1).

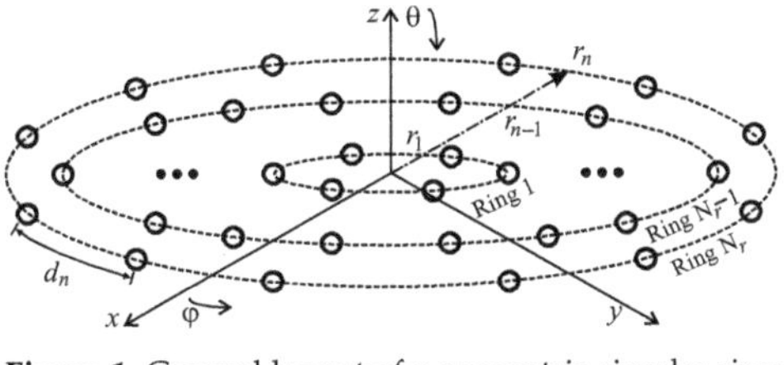

Figure 1. General layout of a concentric circular ring array (CCRA).

Assuming a multi-beam scenario, the array factor is simultaneously optimized with p beams, considering the aperiodicity in function of the different radius in each ring r and an isophoric array feeding [23–25], the AF can be written as follows:

$$AF^p(\theta,\varphi,\vec{r}) = \sum_{n=1}^{N_r}\sum_{m=1}^{N_e} \exp[jkr_n(u\cos\varphi_{nm} + v\sin\varphi_{nm})], \quad (2)$$

In the following section, the theoretical model of the feeding subsystem for the ACRA is described.

2.2. Coherent BFN Model

In this paper, a coherent feeding network with dual-beam shaping and steering capabilities is used in combination with an ACRA in our numeric experimentation. The feeding system relies on the working principles of coherently radiating periodic structures implemented as a beamforming network (also known for its acronym as C-BFN). This feeding network, which combines the redirecting beam control and the power distribution network, uses a unit cell periodically and alternately as a power combiner and divider to lead the signals from the input ports to the radiators. In a C-BFN, each input port obeys an in-phase excitation law (true time delay) and the output ports have a Gaussian-shape amplitude distribution [26,27]. Usually, the topology of the coherent network is built by using unit cells along with consecutive layers, as is illustrated in Figure 2. This specific example shows the signal propagation from three different input ports to the shared outputs. In this feeding system, the power throughout the network shows a binomial excitation behavior [15]. The modular arrangement uses the unit cell as a key piece of the network, facilitating the use of distinct topologies for specific design needs. A C-BFN should guarantee an in-phase performance at the output, no matter its periodicity. Practical research of the feeding system relies on implementing unit cells using different multi-port devices [15,26,28].

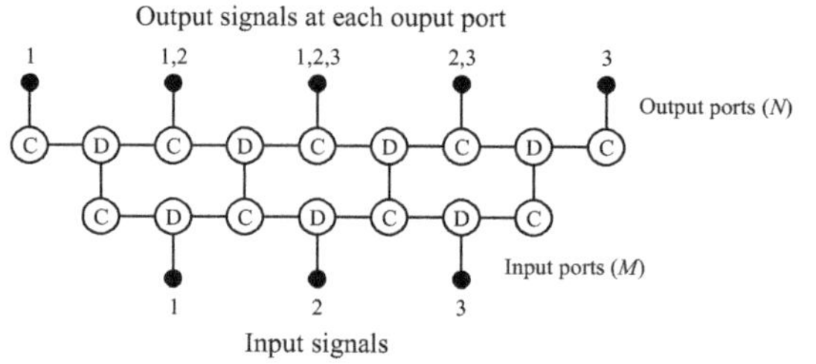

Figure 2. Typical coherent beamforming network (C-BFN) topology based on unit cells for power division/combination, illustrating the shared aperture at the output.

As shown in Figure 2, a three-port component is implemented for power division and combination, and its behavior can be described as follows [15]:

$$[S]_{\text{cell}} = \begin{bmatrix} 0 & j/\sqrt{2} & j/\sqrt{2} \\ j/\sqrt{2} & 0 & 0 \\ j/\sqrt{2} & 0 & 0 \end{bmatrix}. \tag{3}$$

This Wilkinson's matrix ensures isolation among the inputs with a matching behavior that can be observed at the ports. Assuming transmit mode, a transfer matrix can be applied for assessing the final output of the coherent feeding system, as follows [29]:

$$[\vec{\alpha}_{m+1}]_{(m+1,1)} = [T_m]_{(m+1,m)} \cdot [\vec{\alpha}_m]_{(m,1)}, \tag{4}$$

where $[\vec{\alpha}_{m+1}]$ represents the output of the coherent network using one layer. The above can be extended for more layers if necessary. On the other hand, $[\vec{\alpha}_m]$ is the excitation input. In both cases, m represents the feeding ports.

The transfer matrix for the coherent network is expressed as follows:

$$[T_m] = \begin{bmatrix} \sqrt{2}/2 & 0 & 0 & \cdots & 0 \\ 1/2 & 1/2 & 0 & \cdots & 0 \\ 0 & 1/2 & 1/2 & & \vdots \\ 0 & 0 & \ddots & \ddots & 0 \\ \vdots & \vdots & & 1/2 & 1/2 \\ 0 & 0 & \cdots & 0 & \sqrt{2}/2 \end{bmatrix}_{(N,M)}. \tag{5}$$

In the transfer matrix, its coefficients are related to the unit cell, as can be observed in (3), in which the rows denote power combination and the columns represent power division. Consequently, it is possible to synthesize this feeding system with a complete antenna system, constructing an architecture that relies on Figure 2 and evaluating (4) and (5).

2.3. Optimization Tool

This subsection briefly describes the procedure to optimize the two-beam antenna pattern and find adequate solutions for the coherent beamforming network. In this specific case, the task is done by applying a global optimization method and the numerical examples in this paper were implemented using the MATLAB software version R2018a. The optimization tool in this work uses hard constraints and specifications in an objective function, finding candidate solutions to achieve quasi-optimal solutions in the multi-beam scenario. The problem to optimize with constraints is correlated with a complex-valued argument (2) in an exponential function associated with the beam patterns and antenna elements positions, dealing with an oscillating and nonlinear behavior.

In order to find the trade-off between maximum directivity and the lowest peak SLL in a predefined angular location for each beam, the antenna synthesis problem can be expressed as an optimization problem with constraints to be minimized, to obtain a quasi-optimal solution of the radius of each ring in the non-uniform concentric array, in the following way:

$$\begin{aligned} \min_{\vec{r}} \quad & \left\{ \frac{|AF(\theta_{\text{msl}}, \varphi_{\text{msl}}, \vec{r})|}{|AF(\theta_{\max}, \varphi_{\max}, \vec{r})|} \right\} + \left\{ \frac{1}{D(\theta, \varphi, \vec{r})} \right\}, \\ \text{s.t.} \quad & |AF^{p}(\theta_{\max}, \varphi_{\max}, \vec{r})| = 1, \\ & \vec{r} = \{ r_n \,|\, \vec{r} \in \mathbb{R} \}, \; \forall n \in N_r, \end{aligned} \tag{6}$$

where $(\theta_{\text{msl}}, \varphi_{\text{msl}})$ is the direction attained by the peak sidelobes on either side of the main beams, $(\theta_{\max}, \varphi_{\max})$ denote the main beam angles, and the size of the radius domain for each circular ring is indicated by $\mathbb{R}$.

In this case, the optimization process is performed by differential evolution (DE). This algorithm is a proficient and widely used optimizer commonly applied in electromagnetic optimization problems [30]. Summarizing the algorithm in general, the DE optimizer explores for a global optimum solution in a d-dimensional decision space $\Psi \subseteq \Re^d$. It works through an initial population (N_{pop}) with random initialization (within boundary constraints) of parents or real-valued target vectors (genome/chromosomes). Each vector is a candidate solution that enters into an iterative optimization process for the multi-dimensional problem. After the initialization process, in the optimization cycle that contains three different processes, known as mutation, crossover, and selection, some of the candidate solutions are selected and geometrically perturbed. The central component of DE and the principal strategy on the optimization task is carried out by the genetic operator known as mutation. Mutation creates an intermediate mutant/donor vector $\vec{v}_d$ for each vector $\vec{\chi}_i^G = [\chi_{i,1}^G, \chi_{i,2}^G, \ldots, \chi_{i,d}^G]$ in the population at the Gth iteration (called iteration), as follows:

$$\vec{v}_d^{G+1} = \vec{\chi}_{r1}^G + F(\vec{\chi}_{r2}^G - \vec{\chi}_{r3}^G). \tag{7}$$

The scaling factor F indicates a control input (positive) for scaling the difference parameters $(\vec{\chi}_{r2}^G - \vec{\chi}_{r3}^G)$ which defines the length of the exploration, also called the mutation factor. The indices r_1, r_2, and r_3 are mutually exclusive integers randomly taken from the population $[1, N_{pop}]$. Although detailed information about the DE algorithm (in general) can be consulted in [31,32], for sake of simplicity, in Algorithm 1 the pseudocode of the DE scheme used in this work is presented for a better understanding and implementation [33].

The optimization task applied to the antenna system can be summarized as follows: basically, the DE optimizer randomly generates vectors as possible solutions. The candidate vectors of real numbers denote the variables in the optimizer—in this particular optimization case, the radii of the array $\vec{r} = [r_1, r_2, \ldots, r_n]$. Hence, each suggested solution vector proposes a dual-beam radiation pattern to be evaluated that meets certain characteristics of side lobe level and directivity for specific directions of interest in the array. Subsequently, the mutation operator increases the search in the d-dimensional space, followed by the crossover sub-process that handles the components inherited from the mutation operation with a fixed probability crossover rate (Cr), improving the diversity in the population but minimizing convergence speed. Afterwards, the selection operator protects the survival of the fittest solutions in the following generations and reduces the probability of population stagnation. At the end, the algorithm reaches a global best population that conforms a dual-beam power pattern with the predefined design parameters, i.e., a dual-beam pattern with maximum directivity and minimum sidelobe level in a fixed location in our numerical examples.

Algorithm 1: DE/rand/1/bin pseudocode used in the optimization process

```
Input: G: Maximum number of iterations, N_pop: Population size, F: Scale factor,
       Cr: Crossover rate
Output: P_G: Global best solution
Generate I_pop individuals of the initial population pseudo-randomly;
while (G ≠ True) do
    for i = 1 : I_pop do
        Evaluate population f(χ_i)
    end
    for i = 1 : I_pop do
        **Mutation**
        Select 3 individuals χ_r1, χ_r2, χ_r3;
        compute Equation (7)
        **Crossover**
        v_d = χ_off
        for j = 1 : n do
            generate rand(0,1);
            if rand(0,1) < Cr then
                χ_off,j = χ_i,j;
            end
        end
        **Selection**
        if f(χ_off) ≤ f(χ_i) then
            save index for replacement χ_i = χ_off;
        end
    end
    perform replacements;
end
```

3. Numerical Results

In this section, the simulation setup is described alongside the numerical results obtained from the two-beam C-BFN configuration for applications in ACRAs. One main objective under this scenario is to estimate the dual-beam pattern generated by the antenna system (i.e., the beamforming network and the non-uniform antenna array), as well as the performance in a comparative study with a uniform layout scenario in order to compare the results. Following this, we made a comparison with the periodic antenna element distribution using the proposed array feeding network and taking the worst solutions from the aperiodic array. Thus, we evaluate the feasible improvements of non-uniform over the uniform aperture on a CCRA fed by a coherent network with steering and multi-beam capabilities. Additionally, we focused on this from an optimization synthesis problem point of view. The complete multi-beam system is optimized under a fair simulation setup with multiple trials by a global optimizer. The intention behind this consists of, using the obtained behavior of the dual-beam radiation patterns analyzed, showing and exhibiting the complexity reduction and performance improvements of using a two-beam coherent feed system with an ACRA.

Figure 3 illustrates the general representation of the antenna setup utilized. The system shows a CCRA integrated by non-uniform spaced rings. Specifically, without loss of generality, this diagram represents an ACRA constituted by radiant elements (92 in total) located within five different circular rings ($N_r = 5$). The number of radiant elements inside each circular ring in the array is constituted, respectively, by $[N_1, N_2, N_3, N_4, N_5] = [6, 12, 18, 25, 31]$. This radiant element distribution corresponds to $N_e = 2\pi r_n / d_{\min} \mid r_n = n\lambda/2$. In the numerical examples, in order to attenuate any achievable mutual coupling interaction among antenna elements in the array, the spacing between adjacent radiant elements in each circular ring is set at a minimum of $d = 0.5\lambda$ and each circular ring is set to be located by the same minimum distance of 0.5λ with isophoric feeding (a single excitation level) and $r_1 \geq 0.5\lambda$. As depicted in the bottom part of Figure 3, the feeding subsystem is a C-BFN based on Q blocks of one layer, which in total integrates the complete coherent network subsystem. In this considered configuration, the Q required C-BFN blocks per ring is equal to the number of circular rings N_r, i.e., one C-BFN block for every circular ring within the aperiodic array. In this case, the designed configuration for feeding the aperiodic concentric array is constituted by five C-BFN blocks. For the particular case of this proposal, the (N) output ports of each block are interconnected straight to a radiant element in every circular ring, as shown in Figure 3. Furthermore, at the ends of each C-BFN block, the indices denote the amount of output port signals (top of the individual C-BFN blocks) and the input port signals (bottom part of every C-BFN block). In addition, the proposed C-BFN allows for controlling a couple of independent beams simultaneously in the alternation of the feeding ports in each block, as depicted in the lower part of Figure 3. Thus, the number of beams or signals applied is represented by the indices at the feeding ports, i.e. for the conformation of one of the two beams, each index number belongs to a group of complex inputs.

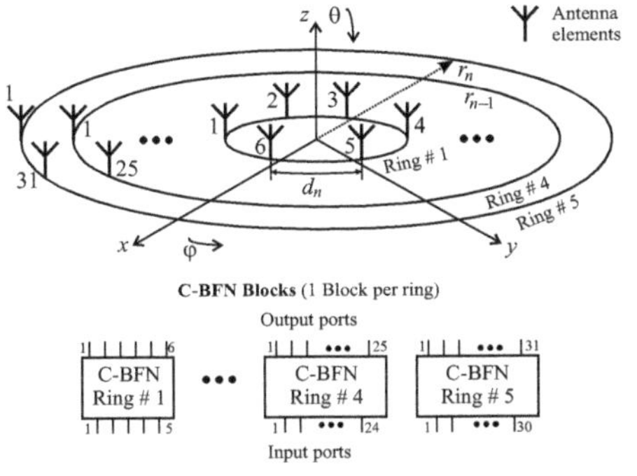

Figure 3. Schematic of the dual-beam array feed network subsystem and the aperiodic antenna array aperture.

For the numerical experiments, the DE algorithm was programmed as well as used to optimize the aperiodic antenna system as well as the feeding network. The evolutionary algorithm is set up to initialize with a population of 120 individuals, a crossover rate CR = 0.9, a factor of mutation fixed in $F = 0.5$, and a maximum number of generations/iterations set in $G_{\max} = 1000$. A random selection, one difference vector, and binomial crossover (DE/rand/1/bin) comprise the DE scheme utilized to optimize the multi-beam radiation pattern. The global stochastic algorithm was run in 20 independent trials for every dual-beam pattern. For this case, the directions of maximum radiation correspond to a slice of the total multi-beam pattern for values of φ (φ-cutting patterns), where the optimized dual-beam array factors are predominately a function of the θ angle.

Figure 4 depicts the performance of the normalized radiation patterns for the two-beam configuration under study (Figure 3) explained in detail in this same section. In these numeric examples, the target of maximum radiation in this multi-beam scenario is steered towards the azimuth plane and directed at 45°, 95°, and 245° for beam number 1. The beam number 2 is set in 345°, 145°, and 295°. The optimization process utilized to improve the radius of every ring $\vec{r} = [r_1, r_2, \ldots, r_5]$ on each dual-beam example allows us to obtain and demonstrate a satisfactory performance regarding isolation level and directivity. Concerning the near-optimal ring spacings (radii), we found that the optimization process in these multi-beam cases achieves the following: $[1.29\lambda, 1.81\lambda, 2.45\lambda, 2.96\lambda, 4.73\lambda]$ for Figure 4a, $[0.98\lambda, 1.61\lambda, 2.24\lambda, 2.98\lambda, 4.26\lambda]$ for Figure 4b, and $[1.45\lambda, 2.14\lambda, 2.66\lambda, 3.28\lambda, 4.85\lambda]$ for Figure 4c, and the aperiodic apertures are illustrated in Figure 5.

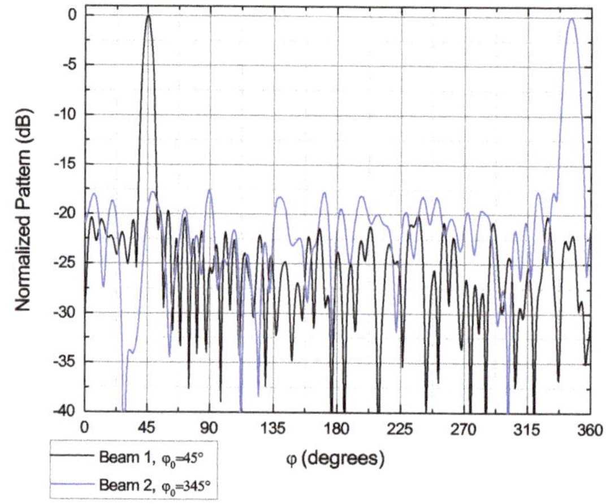

(**a**) Beam number 1 is directed at 45° and beam number 2 at 345°.

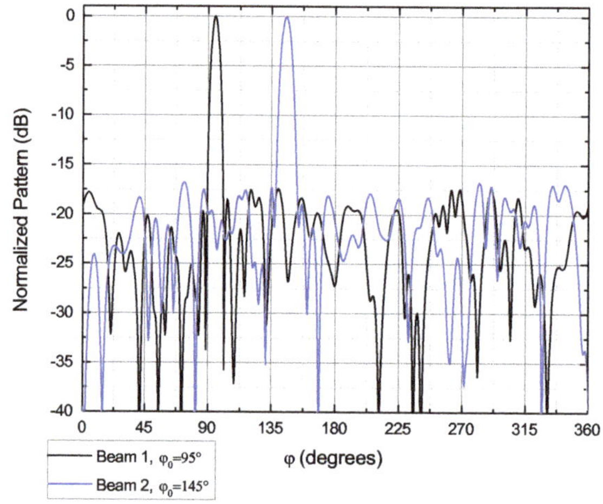

(**b**) Beam number 1 is directed at 95° and beam number 2 at 145°.

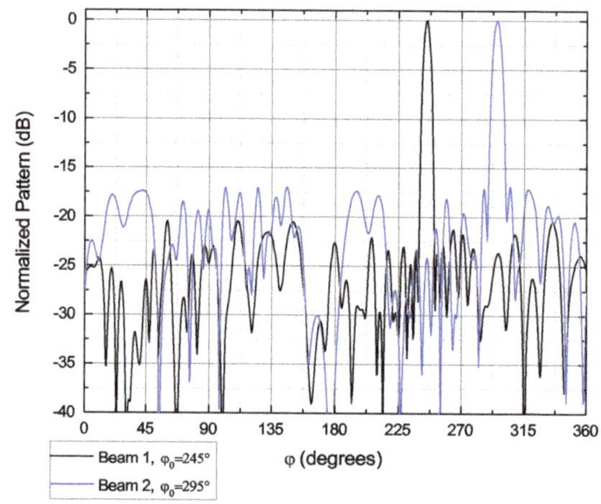

(**c**) Beam number 1 is directed at 245° and beam number 2 at 295°.

Figure 4. Normalized patterns in dB generated by the two-beam feeding network with a non-uniform CCRA optimized using differential evolution.

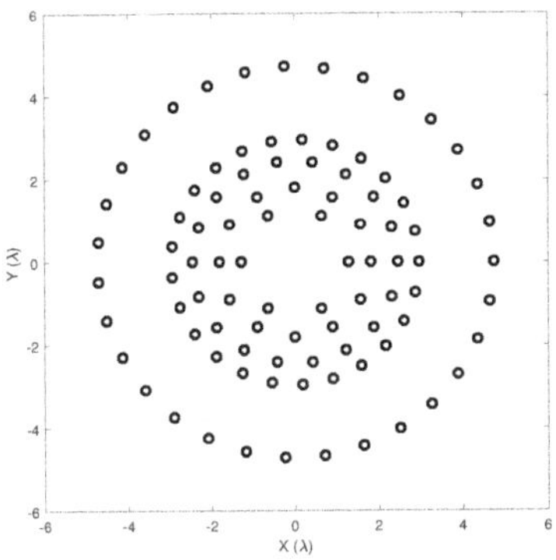

(**a**) Beam number 1 is directed at 45° and beam number 2 at 345°.

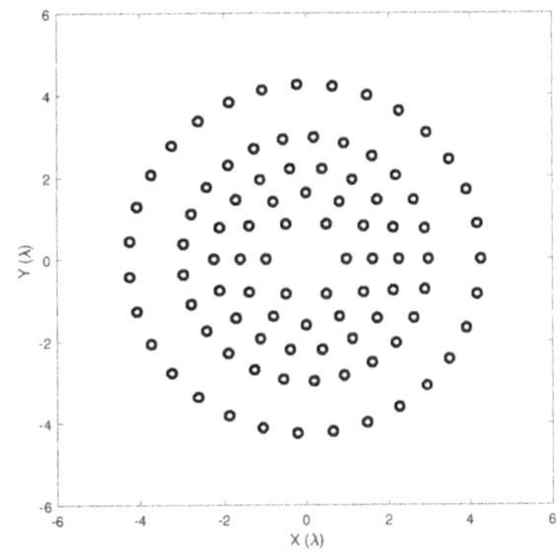

(**b**) Beam number 1 is directed at 95° and beam number 2 at 145°.

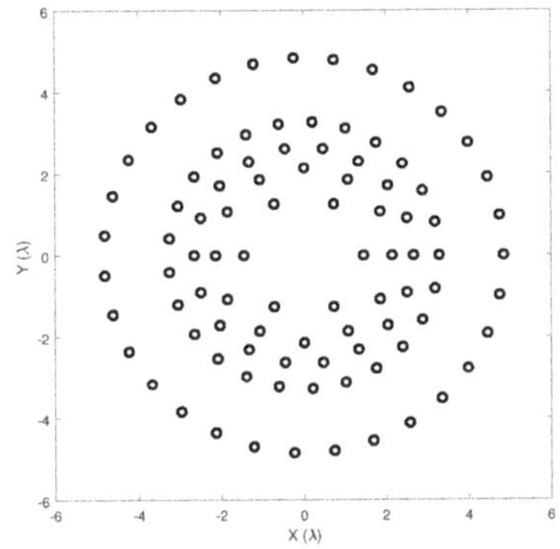

(**c**) Beam number 1 is directed at 245° and beam number 2 at 295°.

Figure 5. Aperiodic layouts of the concentric ring arrays in the numeric examples while optimizing antenna rings ($\vec{r}$).

Furthermore, Table 1 presents the precise numerical values of sidelobe level and directivity retrieved and depicted in each numeric example from Figure 4 with the proposed C-BFN configuration. This table condenses the information from the worst numerical values of SLL and D achieved through the aperiodic aperture via C-BFN on the multi-beam scenario for each numeric example and contrasts the results of the previously mentioned parameters with a uniform aperture (Uniform) with a equivalent coherent feeding network. In addition, Table 1 collects the mean outcomes of the improvement obtained through the aperiodic aperture employing the proposed feeding subsystem on each beam for all of the optimized examples. In each numerical example, the aperiodic layout is straightforwardly set out in comparison with the equivalent uniform layout.

Additionally from Table 1, we can observe that the performance obtained by the worst case of the normalized two-beam patterns generated from the proposed coherent beamforming network outperforms the case of a multiple radiation patterns under the condition of an uniform concentric ring array while using the same feeding system. The general performance shows that a C-BFN configured on a non-uniformly ring-spaced concentric ring array notoriously improves in identical conditions to an uniformly ring-spaced concentric ring array fed by the similar coherent feeding system, in terms of sidelobe level and directivity. The improvement in the shown examples is approximately to −5.6 dB of the sidelobe level and 3.0 dB of the directivity for beam number 1 and −2.5 and 1.3 dB for the same parameters for beam number 2. The improvements (expected values) for each beam are presented in Table 1.

Table 1. Results of the conducted experiments and mean values of the non-uniformity improvement on the proposed antenna system in terms of sidelobe level (SLL) and *D*.

Antenna System (Array + CBFN)	Number of Beam	Main Beam Direction	SLL [dB]	*D* [dB]	ACRA + CBFN Improvement SLL [dB]	*D* [dB]
Uniform	1 and 2	$0 \leq \varphi_0 \leq 360$	−14.49	15.35	–	–
Multi−beam (Opt. $\vec{r}$)	1	45°	−19.67	17.89	−6.2	3.2
		95°	−17.40	17.78	−4.7	2.5
		245°	−20.46	18.88	−6.1	3.5
	2	345°	−17.59	16.12	−3.0	0.9
		145°	−16.74	16.65	−2.2	1.3
		295°	−16.86	17.27	−2.3	1.9

As was demonstrated, the uniform circular concentric geometry fed by a C-BFN maintains the usual nature of circular antenna arrays, retaining inherent high directivity while registering high sidelobe levels over the range of interest. In particular, our proposed two-beam C-BFN configuration that feeds an ACRA characterized by subsets of alternated inputs on each C-BFN block in every ring achieves and retains outstanding numerical values of directivity, i.e., an average performance enhancement between 2.5 and 3.5 dB for beam number 1 and 0.9 and 1.9 for beam number 2 (under equal and fair conditions when running the optimization process). Conversely, all the nonequal layout multi-beam experiments (based on Opt. $\vec{r}$) with the selected coherent system reach higher numerical results of sidelobe level, improving the uniform antenna array, with a performance improvement between −4.7 and −6.2 dB for beam number 1 and −2.2 and −3.0 dB for beam number 2 on average. Furthermore, in this dual-beam configuration, every main beam is shaped and steered to any predefined location utilizing approximately 50% of the input ports (47 input ports for beam number 1 and 45 input ports for beam number 2) for this specific case regarding the total number of radiators ($N = 92$).

Based on the comparative analysis with respect to the dual-beam ACRAs summarized in Table 1. The experimental results prove and confirm that the beamforming network configuration proposal in addition to an ACRA has enhanced performance on each beam compared to a uniform concentric ring array with an equal feeding system regarding the directivity and sidelobe level. Although a predefined direction of maximum radiation is set for each beam in the numerical experiments, the proposed coherent network configuration can generate and scan every beam to any direction of interest beneath the field of view, with a directivity and isolation level with small fluctuations along the azimuthal plane. To establish the maximum performance and for an unbiased analysis in terms of the antenna design trade-off (*D* and SLL), including inherent complexity reduction in the network and multi-beam capability, the improvement of a sparseness optimization on the antenna layout has been avoided. Moreover, the optimized circular rings in Figure 5 are considered to be distanced by at least a separation of $\lambda/2$ plus an additional distance ($0 \leq \Delta_\lambda \leq 1.5\lambda$) among the rings. The novel resulting optimized non-uniform layouts for the multi-beam scenario show a singular comportment where the last and first rings have the biggest radius and with element density in the radial direction; this specific nature is also found in regular non-uniform circular concentric rings and is associated with a low sidelobe amplitude taper.

The designed two-beam feeding system found on C-BFN for ACRA benefits from its operational behavior, with improved features over the conventional circular concentric aperture in a phased array with direct feeding. Its clear benefits include sharing paths coherently (i.e., mechanism of coherent coupling), allowing multi-beam features, and also reducing control signal ports. In the configuration proposed in this paper, the dual-beam is scanned and shaped approximately with half of the input ports, i.e., a clear reduction of 50%, only requiring half of the input ports to control each beam compared to the conventional case.

This novel two-beam configuration approach in the proposed feeding system is not exclusive for aperiodic CCRAs. In fact, a variety of C-BFN setup configurations can be designed by properly modifying the layers or the input/output ports based on the number of beams and the overall performance. Thus, this proposed dual-beam antenna system found on C-BFN with a non-uniform circular concentric aperture is a candidate for different antenna applications considering its isophoric feeding and reduced complexity, as well as its multi-beam shaping performance along with the scanning range.

4. Conclusions

This paper introduced a new dual-beam antenna system for a non-uniform circular concentric ring aperture based on a coherent beamforming network, with scanning and multi-beam shaping capabilities. The global performance of two-beam normalized power patterns, generated by the designed coherent network configuration in an aperiodic and periodic circular concentric ring layout, was contrasted and analyzed. The numerical experiments based on antenna synthesis demonstrate and validate that the designed beamforming network comprises a dual-beam with steering and shaping properties, taking advantage of the layout's aperiodicity, coherent behavior, and isophoric feeding levels, reducing the antenna system's complexity. In the proposed feeding system configuration based on intercalated input signals grouped on C-BFN blocks per ring, numerical results present substantially lower sidelobe levels ($-17.53 \leq \text{SLL} \leq -22.11$ for beam 1 and $-16.92 \leq \text{SLL} \leq -20.61$ on average) and improved directivity values ($17.95 \leq D \leq 19.33$ for beam 1 and $16.47 \leq D \leq 18.73$ for beam 2 on average), showing a significant reduction in the number of control signals (about 50%), i.e., each beam on the two-beam configuration is controlled and scanned with half of input ports, in contrast with the usual feeding system, which requires one feeding input for each output port. A dual-beam coherent network with an aperiodic concentric ring array surpasses its uniform counterpart by up to −6.1 and 3.5 dB on average for the sidelobe level and directivity per beam, respectively. Further studies will explore this antenna system further by using full-wave electromagnetic analysis.

Author Contributions: Conceptualization, A.A.; methodology, A.A. and E.S.-N.; software, A.A. and D.H.C.; validation, A.A., E.S.-N., U.P.-R., M.C.-J., F.R.C.-S. and D.H.C.; formal analysis, A.A., E.S.-N. and D.H.C.; investigation, A.A.; writing—original draft preparation, A.A., E.S.-N. and U.P.-R.; writing—review and editing, A.A., E.S.-N., M.C.-J., F.R.C.-S. and D.H.C.; visualization, A.A., U.P.-R. and M.C.-J.; supervision, A.A., E.S.-N. and D.H.C.; project administration, A.A., and E.S.-N. All authors have read and agreed to the published version of the manuscript.

Funding: This research work was partially funded by the Mexican National Council for Science and Technology (CONACYT) with the project Catedras (No. 872) and the Program for Professors' Development (PRODEP) from the Federal Mexican Government.

Data Availability Statement: The data supporting this research article are available upon request to the corresponding author.

Conflicts of Interest: The authors declare that there is no conflict of interests regarding the publication of this manuscript.

References

1. Blake, J.; Nygren, E.; Schennum, G. Beamforming networks for spacecraft antennas. In *1984 Antennas and Propagation Society International Symposium*; IEEE: Piscataway, NJ, USA, 1984; Volume 22, pp. 158–161. [CrossRef]
2. Anguera, J.; Andújar, A.; Jayasinghe, J. High-Directivity Microstrip Patch Antennas Based on TModd-0 Modes. *IEEE Antennas Wirel. Propag. Lett.* **2020**, *19*, 39–43. [CrossRef]
3. Laue, H.E.A.; du Plessis, W.P. Numerical Optimization of Compressive Array Feed Networks. *IEEE Trans. Antennas Propag.* **2018**, *66*, 3432–3440. [CrossRef]
4. Dorrah, A.H.; Eleftheriades, G.V. Peripherally Excited Phased Array Architecture for Beam Steering with Reduced Number of Active Elements. *IEEE Trans. Antennas Propag.* **2020**, *68*, 1249–1260. [CrossRef]
5. Lin, C.-C.; Tzuang, C.-K.C. A dual-beam micro-CPW leaky-mode antenna. *IEEE Trans. Antennas Propag.* **2000**, *48*, 310–316. [CrossRef]

6. Floc'h, J.M.; El Sayed Ahmad, A.; Kokar, Y. Dual-Beam Antenna Design for Autonomous Sensor Network Applications. *Int. J. Antennas Propag.* **2012**, *2012*, 289681. [CrossRef]
7. Deng, C.; Yu, W.; Lv, X. All-metallic dual-beam antenna with a low profile. In Proceedings of the 2017 IEEE International Symposium on Antennas and Propagation, and USNC/URSI National Radio Science Meeting, San Diego, CA, USA, 9–14 July 2017; pp. 2121–2122. [CrossRef]
8. Hakimi, A.M.; Oraizi, H.; Keivaan, A.; Amini, A. Wideband Vertically Polarized Dual-Beam Antenna Using Modulated Metasurfaces. In Proceedings of the 14th European Conference on Antennas and Propagation (EuCAP 2020), Copenhagen, Denmark, 15–20 March 2020; pp. 4–8. [CrossRef]
9. Xi, R.; Ang, S.; Porter, K.; Li, L. Generation of dual-beam orbital angular momentum vortex beam using transmit arrays. In Proceedings of the 2019 IEEE International Symposium on Antennas and Propagation, and USNC/URSI Radio Science Meeting, Atlanta, GA, USA, 7–12 July 2019; pp. 993–994. [CrossRef]
10. Verevkin, A.P.; Kirillov, V.V.; Munina, I.V.; Turalchuk, P.A. Dual-beam Transmitarray for High Capacity Wireless Communication Systems. In Proceedings of the 2020 IEEE Conference of Russian Young Researchers in Electrical and Electronic Engineering (EIConRus 2020), Moscow, Russia, 27–30 January 2020; pp. 84–86. [CrossRef]
11. Nikfalazar, M.; Sazegar, M.; Mehmood, A.; Wiens, A.; Friederich, A.; Maune, H.; Binder, J.R.; Jakoby, R. Two-Dimensional Beam-Steering Phased-Array Antenna With Compact Tunable Phase Shifter Based on BST Thick Films. *IEEE Antennas Wirel. Propag. Lett.* **2017**, *16*, 585–588. [CrossRef]
12. Liu, S.; Chen, Q. Dual-Beam Gain-Reconfigurable Antennas Using A Shared Reflectarray Aperture. In Proceedings of the 2019 International Symposium on Antennas and Propagation (ISAP)-Proceedings, Xi'an, China, 27–30 October 2019; pp. 1–3.
13. Chou, H.; Jian, H. Numerical design of dual-beam open-end waveguide antenna array for the angular diversity at millimeter waves. In Proceedings of the 2016 Asia-Pacific International Symposium on Electromagnetic Compatibility (APEMC 2016), Shenzhen, China, 17–21 May 2016; Volume 1, pp. 118–120. [CrossRef]
14. Kodgirwar, V.P.; Deosarkar, S.B.; Joshi, K.R. Design of Dual-Band Beam Switching Array for Adaptive Antenna Applications Using Hybrid Directional Coupler and E-Shape Slot Radiator. *Wirel. Pers. Commun.* **2020**, *113*, 423–437. [CrossRef]
15. Betancourt, D.; del Rio Bocio, C. A Novel Methodology to Feed Phased Array Antennas. *IEEE Trans. Antennas Propag.* **2007**, *55*, 2489–2494. [CrossRef]
16. Betancourt, D.; del Rio, C. 3 by 3 phased array controlled by only three phase shifters. In Proceedings of the 2006 First European Conference on Antennas and Propagation, Nice, France, 6–10 November 2006; pp. 1–3. [CrossRef]
17. Panduro, M.A.; del Rio-Bocio, C. Design of beam-forming networks using CORPS and evolutionary optimization. *AEU Int. J. Electron. Commun.* **2009**, *63*, 353–365. [CrossRef]
18. Arce, A.; Covarrubias, D.H.; Panduro, M.A. Design of a multiple-beam forming network using CORPS optimized for cellular systems. *AEU Int. J. Electron. Commun.* **2012**, *66*, 349–356. [CrossRef]
19. Saliani, N.; Bemani, M.; Kazemi, H. Modified CORPS beamforming network with null-steering capability. *IET Microw. Antennas Propag.* **2019**, *13*, 1208–1213. [CrossRef]
20. Arce, A.; Stevens-Navarro, E.; Cardenas-Juarez, M.; Pineda-Rico, U.; Covarrubias, D.H. A Coherent Multiple Beamforming Network for a Non-uniform Circular Antenna Array. *Radioengineering* **2019**, *27*, 74–83. [CrossRef]
21. Arce, A.; Stevens-Navarro, E.; Cardenas-Juarez, M.; Pineda-Rico, U.; Simon, J.; Panduro, M.A. Design and Optimization of a Coherent Beamforming Network for an Aperiodic Concentric Ring Array. *Int. J. Antennas Propag.* **2019**, *2019*, 1–10. [CrossRef]
22. Balanis, C.A. *Antenna Theory: Analysis and Design*, 3rd ed.; John Wiley & Sons: New York, NY, USA, 2005.
23. Haupt, R.L. Optimized element spacing for low sidelobe concentric ring arrays. *IEEE Trans. Antennas Propag.* **2008**, *56*, 266–268. [CrossRef]
24. Salas-Sanchez, A.A.; Fondevila-Gomez, J.; Rodriguez-Gonzalez, J.A.; Ares-Pena, F.J. Parametric Synthesis of Well-Scanning Isophoric Pencil Beams. *IEEE Trans. Antennas Propag.* **2017**, *65*, 1422–1427. [CrossRef]
25. Bencivenni, C.; Ivashina, M.V.; Maaskant, R. Reconfigurable aperiodic array synthesis by Compressive Sensing. In Proceedings of the 2016 10th European Conference on Antennas and Propagation (EuCAP), Davos, Switzerland, 10–15 April 2016; pp. 1–3. [CrossRef]
26. Fonseca, N.J.G. Design and implementation of a closed cylindrical BFN-fed circular array antenna for multiple-beam coverage in Azimuth. *IEEE Trans. Antennas Propag.* **2012**, *60*, 863–869. [CrossRef]
27. Rotman, R.; Tur, M.; Yaron, L. True Time Delay in Phased Arrays. *Proc. IEEE* **2016**, *104*, 504–518. [CrossRef]
28. Zaker, R.; Abdipour, A.; Tavakoli, A. Full-wave simulation, design and implementation of a new combination of antenna array feed network integrated in low profile microstrip technology. *Analog. Integr. Circuits Signal Process.* **2014**, *80*, 507–517. [CrossRef]
29. Ferrando, N.; Fonseca, N.J.G. Investigations on the efficiency of array fed coherently radiating periodic structure beam forming networks. *IEEE Trans. Antennas Propag.* **2011**, *59*, 493–502. [CrossRef]
30. Deb, A.; Roy, J.S.; Gupta, B. A Differential Evolution Performance Comparison: Comparing How Various Differential Evolution Algorithms Perform in Designing Microstrip Antennas and Arrays. *IEEE Antennas Propag. Mag.* **2018**, *60*, 51–61. [CrossRef]
31. Eltaeib, T.; Mahmood, A. Differential evolution: A survey and analysis. *Appl. Sci.* **2018**, *8*, 1945. [CrossRef]
32. Opara, K.R.; Arabas, J. Differential Evolution: A survey of theoretical analyses. *Swarm Evol. Comput.* **2019**, *44*, 546–558. [CrossRef]
33. Neri, F.; Tirronen, V. Recent advances in differential evolution: a survey and experimental analysis. *Artif. Intell. Rev.* **2010**, *33*, 61–106. [CrossRef]

applied sciences

MDPI

Article

Maple-Leaf Shaped Broadband Optical Nano-Antenna with Hybrid Plasmonic Feed for Nano-Photonic Applications

Inzamam Ahmad [1], Shakir Ullah [1,*], Jalal ud din [2], Sadiq Ullah [1], Waseem Ullah [1], Usman Habib [3,*], Salahuddin Khan [4] and Jaume Anguera [5]

1 Department of Telecommunication Engineering, University of Engineering and Technology, Mardan 23200, KP, Pakistan; inziahmad950@gmail.com (I.A.); sadiqullah@uetmardan.edu.pk (S.U.); waseemu402@gmail.com (W.U.)
2 Department of Electrical Engineering, University of Engineering and Technology, Peshawar 25120, KP, Pakistan; jalal6422@gmail.com
3 School of Computer Science and Electronic Engineering, Bangor University, Bangor LL57 2DG, UK
4 College of Engineering, King Saud University, P.O. Box 800, Riyadh 11421, Saudi Arabia; drskhan@ksu.edu.sa
5 Electronics and Telecommunication Department, Universitat Ramon Llull, 08022 Barcelona, Spain; jaume.anguera@salle.url.edu
* Correspondence: shakirhayat.eng@gmail.com (S.U.); u.habib@bangor.ac.uk (U.H.)

Abstract: This article presents a broadband optical nano-antenna, which covers a broader range of optical communication wavelengths (666 to 6000 nm), used in nano-photonic applications. The proposed design is modeled and analyzed to obtain a satisfactory gain of up to 11.4 dBi for a single element-based antenna. The unique feature of the proposed antenna is the hybrid plasmonic waveguide-based feed, which receives the optical signal from the planar waveguide and redirects the signal out of the plane. The proposed antenna provides highly directional radiation properties, which makes it a suitable candidate for inter- and intra-chip optical communications and sensing applications. Moreover, an extension of the work is performed for an array configuration of the order 2 × 1 and 64 × 1, to increase the gain and directionality. Therefore, this shows that it can be equally useful for optical energy harvesting applications with a significant gain up to 26.8 dBi.

Keywords: optical nano-antenna; nano-photonic; on-chip optical communication; optical energy harvesting

Citation: Ahmad, I.; Ullah, S.; din, J.u.; Ullah, S.; Ullah, W.; Habib, U.; Khan, S.; Anguera, J. Maple-Leaf Shaped Broadband Optical Nano-Antenna with Hybrid Plasmonic Feed for Nano-Photonic Applications. *Appl. Sci.* **2021**, *11*, 8893. https://doi.org/10.3390/app11198893

Academic Editor: Anming Hu

Received: 24 August 2021
Accepted: 16 September 2021
Published: 24 September 2021

1. Introduction

Antennas have been a significant component for various microwave applications and recent applications of antennas involve transmission in the near-infrared and optical regions. These applications can offer multiple advantages with improved operating conditions, emission control, and light wave radiation into free space [1]. Nature-inspired designs have brought tremendous technological advancements for various applications in recent decades. Antenna geometries based on bio-inspired shapes have been investigated in recent literature, where a flower shape is used to obtain a large bandwidth [2] or a papaya leaf shaped antenna provides a multi-band operation [3]. In terms of antenna array, a four-leaf based antenna is also explored [4] for low frequency applications. The usefulness of the bio-inspired antenna shape for RF applications has been extended to the GHz band [5], where a vine leaf-based shape is designed on a FR4 substrate. This work explores a similar shape (maple-leaf) for THz photonic applications and investigates it for various design parameters and antenna-array configurations. Nature-inspired designs have shown their feasibility towards a variety of significant frequency ranges. A jasmine flower shaped monopole antenna is presented in [6], which provides a large bandwidth in the sub-6 GHz range frequencies. A triangularis plant's leaf for WiFi frequency range (2.4 GHz band) is simulated and fabricated in [7] and a snowflake structure for 5G applications is shown in [8]. The use of Fractal geometry to design these shapes is explored in [9], which is used by [10] to design a Spiral shell shaped antenna for the multi-band frequency operation. Additionally, it is extended to a dielectric resonator based prototype design in [11].

The optical antenna is a practicable application of physical optics that has the capability of capturing and converting free propagating optical electromagnetic radiation to localized energy and vice versa [12]. These antennas resemble the typical microwave counterparts as they deal with optical fields to confine, transmit, and receive. By making use of the localized surface plasmons (LSP), these antennas can restrain optical (light) waves into very small volumes [13]. The basic principle for that is the induction of free elections by the metal particles due to the illumination of light. These electrons oscillate as a result and, in turn, generate surface charges which change corresponding to the incident light. The aggregations of surface charges that happen on opposite sides of the particles are called surface plasmons (SPs) and they are actually coherent electron oscillations that arise when there is an interface between metal and dielectric [14]. Although optical antennas are strongly analogous to their radio-frequency (RF) and microwave counterparts, their physical properties and scalable behaviors are much different. Most of these differences arise due to the fact that metals are not perfect conductors [15] at optical frequencies. However, they are strongly correlated plasmas, and thus can be described as a free electron gas that can be defined and solved using Drude model equations [16]. For these reasons, modern advancements in nanotechnology and plasmonic technology have generated substantial interest in the optical antenna concept to efficiently overcome the diffraction limit and make it possible to manipulate optical fields on a nanometer scale [17]. The optical nano-antennas can impressively increase the light-matter interaction efficiency, to get a wide range of applications, such as chemical and biological sensing [18], high-resolution microscopy [19], and nanoscale imaging [20]. Additionally, nano-antennas have been considered as efficient and promising elements in photo-detection [21], heat-transfer (thermal ablation of liver and cancer cell) [22], inter-and/or intra-chip optical communications or radio-over-fiber transmission [23], spectroscopy [24], and energy harvesting [25,26].

Several different nano-antenna configurations have been demonstrated so far, such as bow-ties [27], Yagi-Uda [28], cross antennas [29], half-moon shape antenna [30], Crescents [31], and Nanorods [32]. Most of them are with limited bandwidth due to their dipolar type of resonances. Therefore, these configurations are only suitable for narrowband applications. Optical antennas are required to support the broadband operation for a variety of applications. Optical broadband antennas are also exceedingly desirable for fluorescence enhancement, surface-enhanced Raman scattering, and higher harmonic generation, which are intrinsically multi-wavelength and broadband in nature [33]. However, broadband antennas operating in optical frequencies experience limitations such as low radiation efficiency or narrow frequency band operation, etc. [34]. This problem needs to be addressed with an efficient design of a broadband antenna that can provide radiation coverage to the three main frequency windows for optical transmission (193.1, 229, and 352.9 THz) to be compatible with commercial nano-photonic applications. In order to be able to receive and transmit optical and infrared signals, a fundamental requirement is to feed the antenna with plasmonic nano-guiding. The wedge plasmon waveguides are superior as compared to groove waveguide structures, due to their strong subwavelength localization of guided plasmonic signals, large propagation distances, and relatively low dissipation [35]. Different types of metallic nanostructures have been proposed for guiding SPP modes, including thin metal films (gold, silver) [36], chains of metal nanoparticles [37], sharp metal wedges [38], etc. The mathematical model of bio-inspired shapes can be obtained using the Gielis formula [39] by transforming multiple superellipses to a leaf or flower shape. The key factor is the use of polar coordinates, along with the use of rotational symmetry to obtain a set of six parameters, which can provide a variety of bio-inspired geometries.

In this manuscript, we propose a broadband optical nano-antenna to achieve high performance in terms of broadband transmission characteristics, fed with a hybrid plasmonic waveguide, and structured according to the nature and landscape-inspired shape of a maple leaf. The major contribution of this work is the characteristic design to achieve a high gain at the THz range and the feasibility to increase it in an array configuration. The bio-inspired shape consists of multiple sharp edges on the radiating patch, which helps the antenna structure resonate on multiple frequencies, thus providing a multi-band operation or obtaining a large bandwidth by covering a range of optical frequencies. The designed antenna in this work has been impedance matched with the characteristic impedance of the waveguide feed and can cover all of the optical communication wavelengths (850, 1310, and 1550 nm). The proposed antenna can be used in inter- and intra-chip optical communications and sensing applications. Furthermore, an analysis on the impact of using an array of these designs with 2 × 1 and 64 × 1 configurations are also presented to achieve a highly directive pattern of 15 dBi gain, in order to enable the designed antenna for the application of energy harvesting.

The rest of the paper is planned in the following order. Section 2 describes the theory of hybrid plasmonic waveguides. Section 3 is concerned with designing the procedure and the geometry of the maple shape nano-antenna. Section 4 presents the simulation results and Section 5 concludes the work.

2. Structure

2.1. Modeling and Characterization of Hybrid Plasmonic Waveguide

The designed antenna is based on the hybrid plasmonic structure. Figure 1a shows the basic geometry of the waveguide and standpoint view of the final design. Figure 1b presents the TM mode of the waveguide. These structures are configured by implementing a material with a low refractive index (such as SiO_2), between a metal (Ag) and another material with a higher refractive index (Si). Plasmonic TM modes are the supporting mode, having high confinement inside a low refractive index material [40]. Fundamentally, the hybrid plasmonic waveguide (HPWG) is the combination of dielectric and plasmonic waveguides. Although dielectric waveguides can be considered practically lossless, the mode size in such guides can be limited by the amount of diffraction. On the other hand, plasmonic guides can squeeze light far below the diffraction limit, but consequently cause a large propagation loss. By combining both of these guides, an HPWG can be formed. As compared to purely plasmonic waveguides, HPWG suggests a better compromise between loss and confinement, compatible with silicon on the insulator technology [41]. Different types of waveguides have been analyzed recently. Among these, a hybrid dielectric-loaded plasmonic waveguide (HDLPW) has been proposed as a potential candidate for the optical transmission and its waveguide properties were numerically analyzed to verify the principle technique [42]. In [43], the plasmonic slot nano-antenna is shown to provide a 100 times smaller size than the area of the diffraction-limited spot, to control the optical fields at nano-scale.

2.2. Design of the Nano-Antenna

The geometry of the proposed novel broadband optical nano-antenna that operates in the infrared and visible window is depicted in Figure 2, in terms of the front and side outlook. The design consists of a landscape-inspired Canadian flag-like antenna of length and width represented as L and W, respectively. The antenna is fed by a hybrid plasmonic waveguide which achieves a strong light confinement by coupling the guided light from the dielectric waveguide and plasmonic waveguide. The side view of the antenna in Figure 2b shows that the design comprises three consecutive layers namely metal, low refractive index material, and dielectric having a high refractive index. The low refractive index material is sandwiched between metal and dielectric. The parametric values for the different dimensions shown in the cross-section view of the waveguide are as follows: The thickness of the metallic (Ag) layer, thin SiO_2 layer, and the Si layer are defined as h_{Ag} = 100 nm, h_{SiO2} = 10 nm, and h_{si} = 100 nm, respectively. The height of the substrate (h) is kept as 200 nm. To cover all of the optical communication windows, we fix our target frequency range of operation as 50–450 THz with the center frequency at fc = 250 THz (λc = 1200 nm). The length (*L*) and width (*W*) of the antenna can be calculated using Equations (1)–(6) (where the effective permittivity of Ag is (−129 + 3.28 j) [38]. Table 1 represents the parametric values used for the simulations to obtain optimum performance of the designed antenna. The impedance of the waveguide is determined as follows:

$$Z_{waveguide} = \frac{Z_0}{n_{eff}} = \frac{377}{n_{eff}} \tag{1}$$

where the effective refractive index (n_{eff}) depends on the permittivity of the medium (ε_m) and dielectric (ε_m), as well as, the function of width (*W*) and thickness of SiO_2 (h_{SiO2}). Z_0 is the impedance of the free-space (vacuum), equal to 377 Ω.

$$n_{eff} = \sqrt{\frac{\varepsilon_m \varepsilon_d}{\varepsilon_m + \varepsilon_d}} F(W, h) \tag{2}$$

$$F(W, h_{SiO_2}) = \frac{1 + 1798W + 311h_{SiO_2} + 10.55Wh_{SiO_2}}{32,300 + 759.7W + 3337h_{SiO_2} + 10.55Wh_{SiO_2}} \tag{3}$$

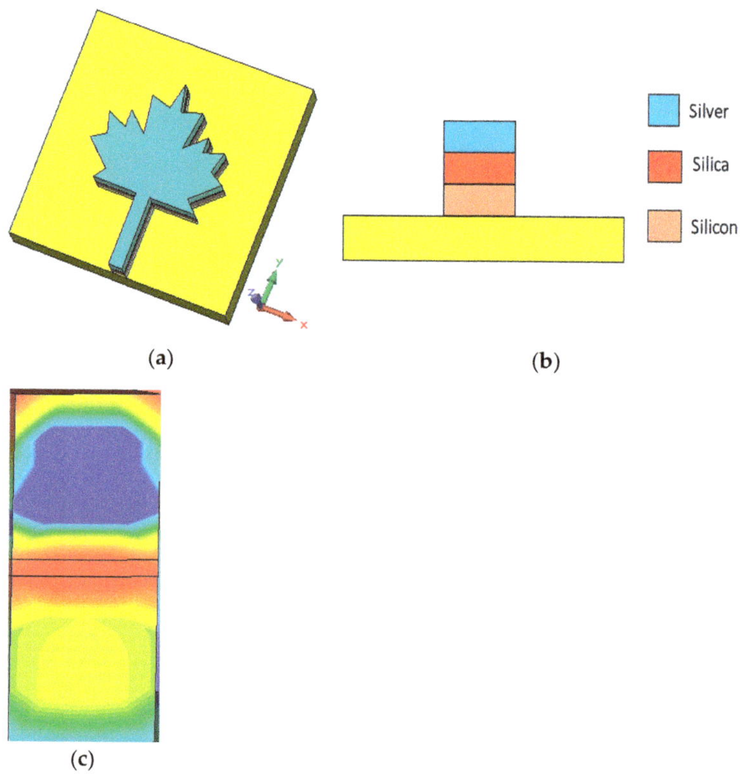

Figure 1. (**a**) Perspective view of the structure; (**b**) basic geometry of HPWG; and (**c**) TM mode.

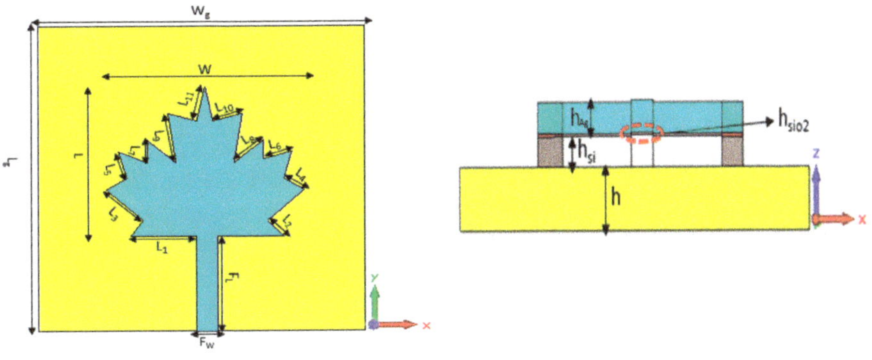

(a) (b)

Figure 2. (**a**) Front view; (**b**) side view.

Table 1. Geometric parameters of the proposed design.

Parameters	Values (nm)	Parameters	Values (nm)
W	960	L_6	155
L	780	L_7	125
W_g	1600	L_8	200
L_g	1600	L_9	280
F_L	500	L_{10}	155
F_W	100	L_{11}	200
L_1	380	h_{Ag}	100
L_2	115	his	100
L_3	250	h_{SiO2}	10
L_4	125	Hg	200
L_5	155	D	600
W_A	1600	L_A	3160

Finally, the antenna impedance, in terms of conductance (G) can be found as:

$$Z_{antenna} = \frac{1}{Y_{in}} = \frac{1}{2G} \quad (4)$$

$$G = \frac{W}{120\lambda_0}\left[1 - \frac{1}{24}\left(\frac{2\pi h_{SiO_2}}{\lambda_0}\right)^2\right] \quad (5)$$

where λ_0 is the free-space wavelength. The effective resonant length of the radiating patch is given by:

$$L = \frac{\lambda_0}{2n_{eff}} \quad (6)$$

The development of the design is performed in the steps shown in Figure 3, which produce a different set of return loss characteristics, as will be discussed in the next section. Changes have been made for an improvement in the antenna performance in terms of the driving point impedance bandwidth and the return loss. Figure 3 depicts that step 1 is designed by introducing a rectangular rod, having two triangular cuts on top of the rod with an operating band of 350 and 450 THz. In step 2, the patch is embedded with the same-sized rod with a wide band of operating frequency of 50–450 THz. A complete maple-leaf shape is achieved with better results in step 3 having a broadband of (50–450 THz) with the resonance of 193, 229, 353, and 390 THz.

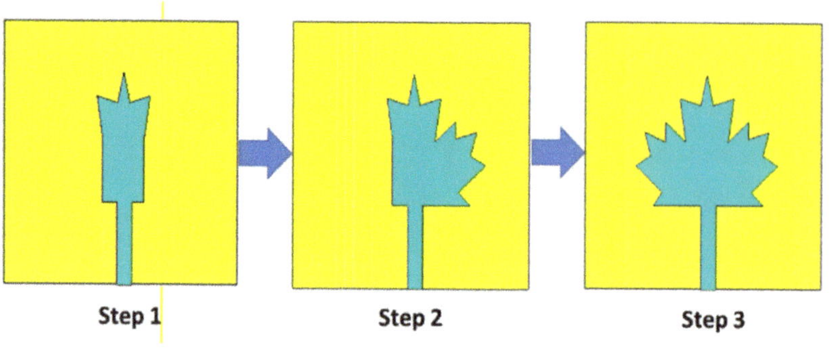

Figure 3. Varied design steps of the proposed antenna.

3. Simulation and Results

For a broadband nano-antenna that can operate in the optical and infrared regions to radiate the optical signal into the free space [44], the presented maple-leaf shape nano-antenna is a promising solution. The proposed nano-antenna has been designed and analyzed using the CST Microwave Studio software package [45], which is based on the technique of the finite integration (FI) solver. The return loss for the antenna design at each step (shown in Figure 3) is plotted in Figure 4.

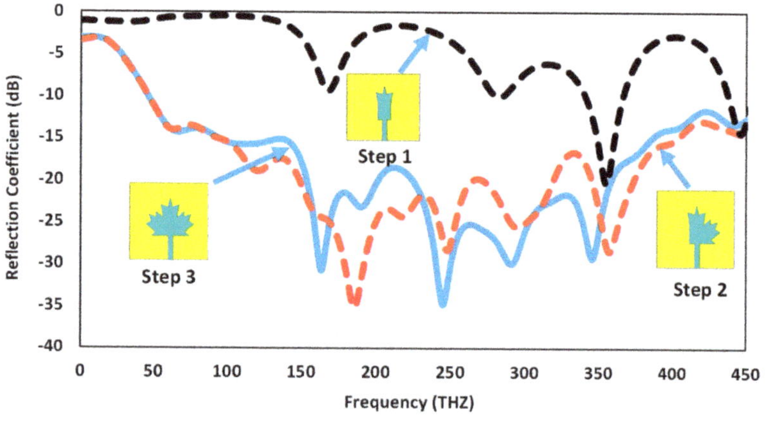

Figure 4. The S_{11} of the design evaluation at each step.

The return loss curve and gain for the final design are plotted against the operational frequency in Figure 5. Here, it can be seen that the antenna operates in all of the three main windows for optical frequencies between 50 to 450 THz with a return loss less than −10 dB. The second observation from the results is that the gain increases with the frequency, which is very useful to compensate for the high propagation loss at higher frequencies. The voltage standing wave ratio (VSWR) of the proposed broadband nano-antenna is depicted in Figure 6. The VSWR value is between 1 and 1.5 for the frequency band of 50 to 450 THz, which is an acceptable value for an antenna operating at these high frequencies. The VSWR value is found to be 1.15, 1.18, 1.14, and 1.4 at 193, 229, 353, and 390 THz, respectively. The maple-shaped antenna is evaluated for different lengths of the leaf tips, as shown in Figure 7. The shape with small leaf tips resonates in multiple frequency bands with the reduced bandwidth. The increase in the leaf tips of a common maple-shape provides better results, but as the length of the leaf tips increases, the dimensions become very narrow and the proposed design will be extremely difficult to be used for real applications, especially in a large array where such small structure details are impossible to achieve.

The simulation work has been conducted by considering the limitations of the fabrication processes. Therefore, a feasible design is selected for the further analysis.

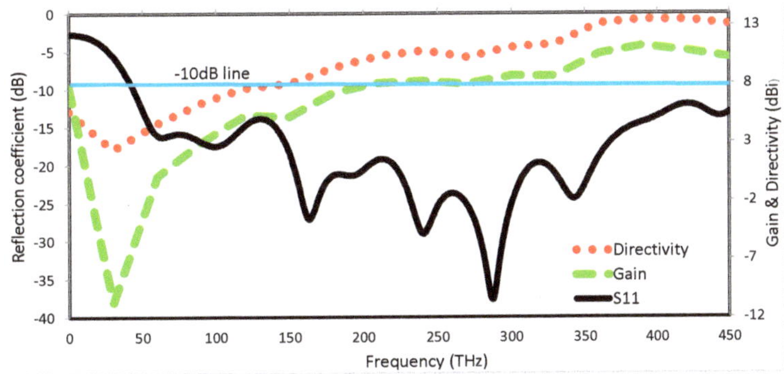

Figure 5. S_{11} and gain of the antenna versus frequency in THz covering all of the three main optical windows.

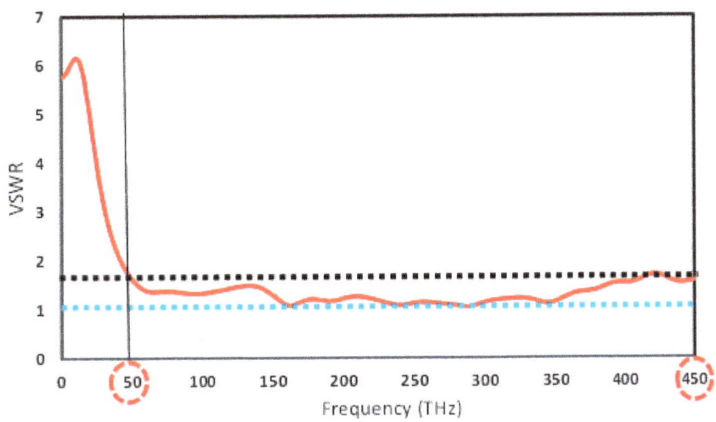

Figure 6. VSWR values for the proposed antenna.

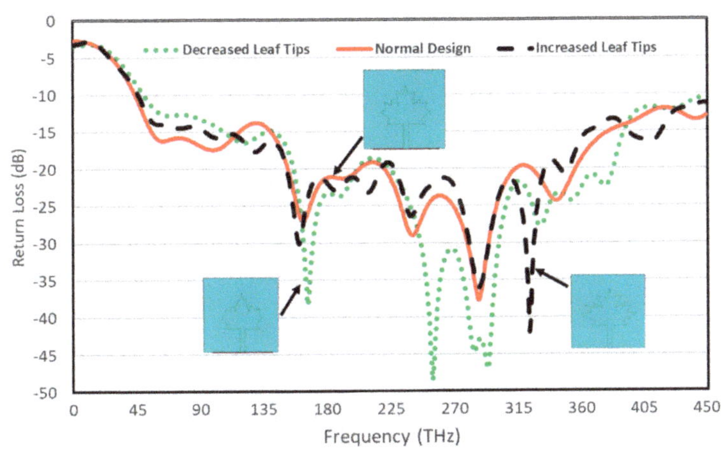

Figure 7. Performance comparison for different lengths of the leaf tips with the maple-leaf antenna geometry shown as an inset.

The polar plot and 3D radiation pattern for the simulated broadband nano-antenna are shown in Figures 8 and 9, respectively. The simulation results show that the proposed optical antenna directs the optical signal at 0° with the half-power beamwidth (HPBW) of 49.5, 41.4, 32.8, and 26.4°, while operating at a frequency of 193, 229, 353, and 390 THz, respectively. From these values, it can be concluded that as the transmitting frequency increases, the beamwidth of the antenna narrows down, and the antenna becomes more directive.

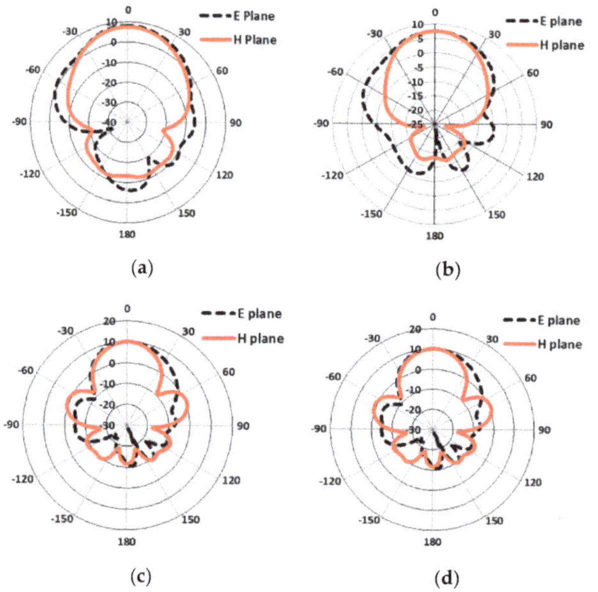

Figure 8. Gain plot in the E and H planes at (**a**) 193 THz; (**b**) 229 THz; (**c**) 353 THz; (**d**) 390 THz.

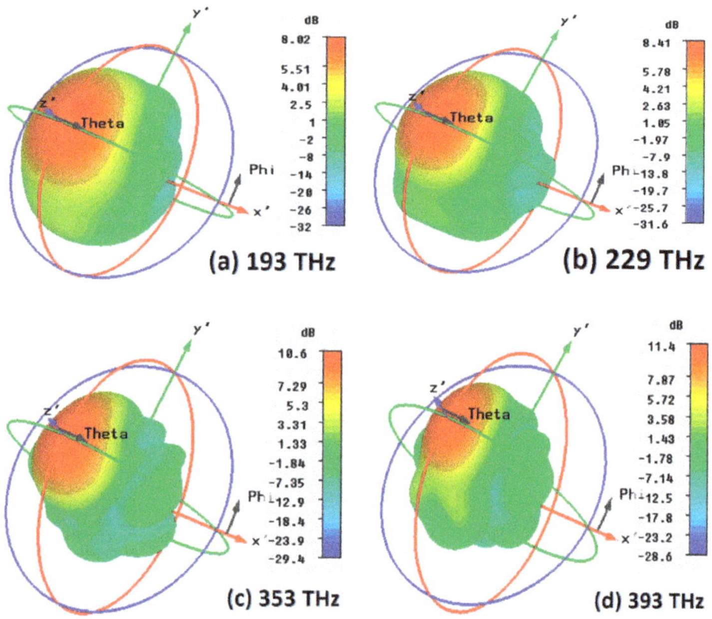

Figure 9. Three-dimensional far-field gain pattern for the designed antenna at (**a**) 193 THz (**b**) 229 THz (**c**) 353 THz and (**d**) 393 THz

To observe the resonant area, surface electric fields (E field) recorded at the resonance frequency of 193 THz are indicated in Figure 10, where the upper and lower edge of the maple-leaf shaped broadband antenna show more intensity than the central region. The effective resonant area is highlighted in Figure 11 for the surface current distributions of the proposed nano-antenna. As expected, at the resonance frequency of 193 THz, most of the current density is concentrated at the central area of the antenna.

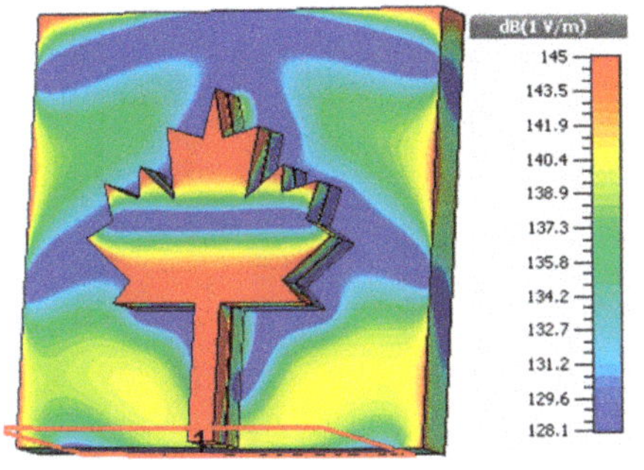

Figure 10. Electric field surface plot of the maple-shaped nano-antenna.

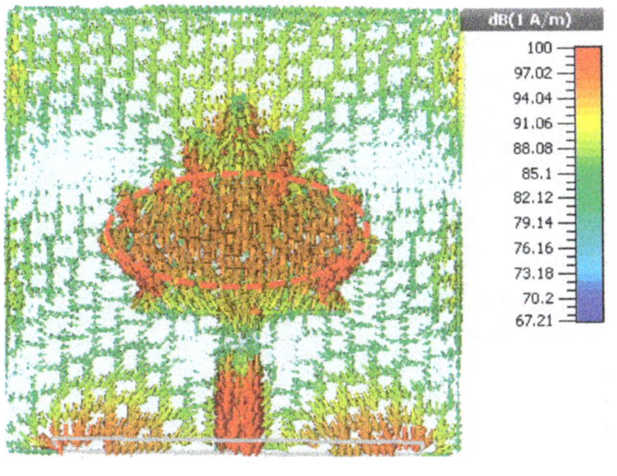

Figure 11. Surface current plot for the maple shape.

4. Antenna Array for Energy Harvesting Applications

An extraordinary amount of work has been done in the field of energy harvesting applications through antennas in the last decade, to minimize the huge consumption of fossil fuel [46]. For energy harvesting applications, the antenna must have a high gain and radiation efficiency, which can be attained by making an array of several antennas [47]. The idea of collecting solar energy from the sun's and earth's radiation using nano-antennas is based on the fact that when a light source impinges the surface of a nano-antenna, the surface plasmons resonate and a time-varying current is induced on the antenna's surface. This phenomenon generates a voltage at the feeding point of the antenna. The induced electric field is then transferred via a feeding line and converted into an electric current to be supplied to a load [48]. By designing a 2×1 array from the aforementioned nano-antenna, a good amount of improvement can be achieved in terms of gain and directivity. The array design for a basic 2×1 configuration is shown in Figure 12, where for a single array with multiple elements, the spacing D is an important parameter that defines the main behavior of the array. The return loss and gain of a single element and 2×1 array are plotted in Figure 13 for comparison. The figure shows that increasing the elements in the array provides a higher gain, but the frequency response of the antenna slightly changes due to the dependence on the value of D. However, both configurations work in the large optical frequency span and can be considered for energy harvesting applications.

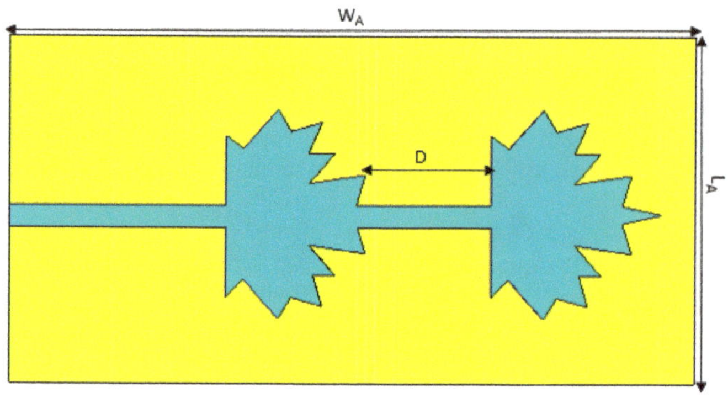

Figure 12. Front view for the 2 × 1 antenna array.

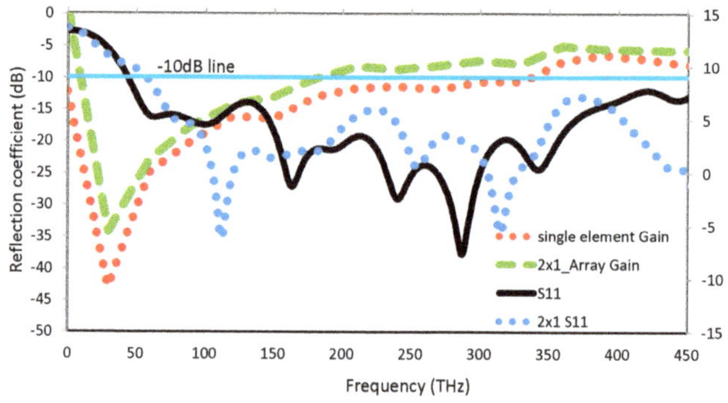

Figure 13. S_{11} and gain comparison of the single element and 2 × 1 array.

As the spacing between the two array elements needs to be carefully optimized, an analysis is performed to show the impact on the return loss. Figure 14 shows the change in response of the designed antenna for different antenna spacing values (D). For a maximum transmission frequency of 450 THz, the results shown previously in Figure 13 are plotted when D = 600 nm is approximately equal to λ. The results for D = λ/2 show a poor return loss and an increase in D to a value of 1.5 λ shows a very low bandwidth operation, which is not suitable for THz applications. Therefore, D of 600 nm is selected for the further analysis and Figure 15 shows the radiation efficiency for the 2 × 1 array. The polar plot and 3D radiation graphs of the 2 × 1 array are shown in Figures 16 and 17, respectively.

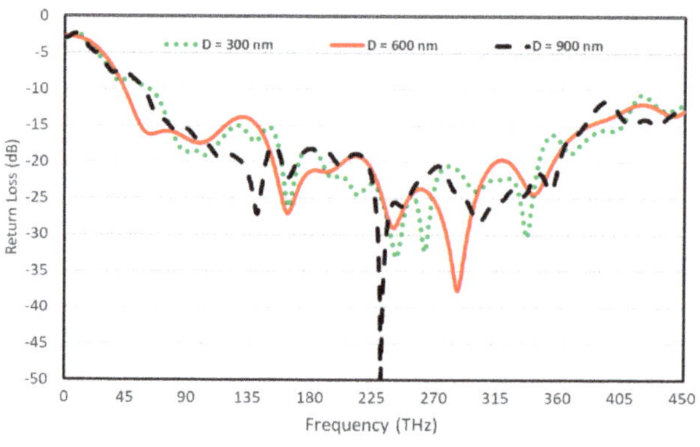

Figure 14. Performance comparison for various spacings between the elements in the 2 × 1 array design.

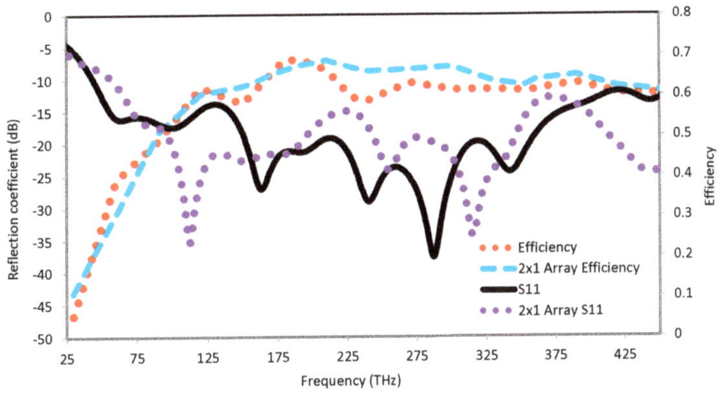

Figure 15. Radiation efficiency for the 2 × 1 antenna array.

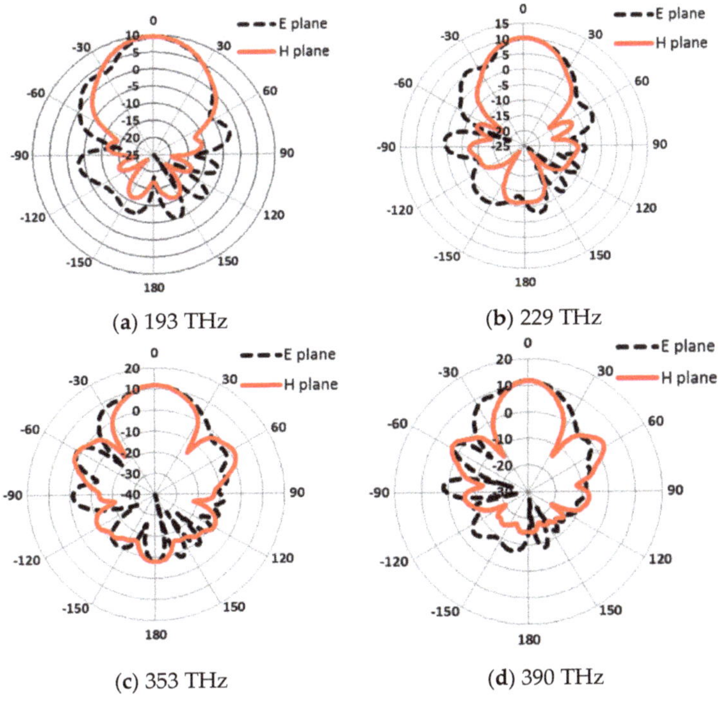

Figure 16. The 2 × 1 array gain plots in the E and H planes at (**a**) 193 THz; (**b**) 229 THz; (**c**) 353 THz; (**d**) 390 THz.

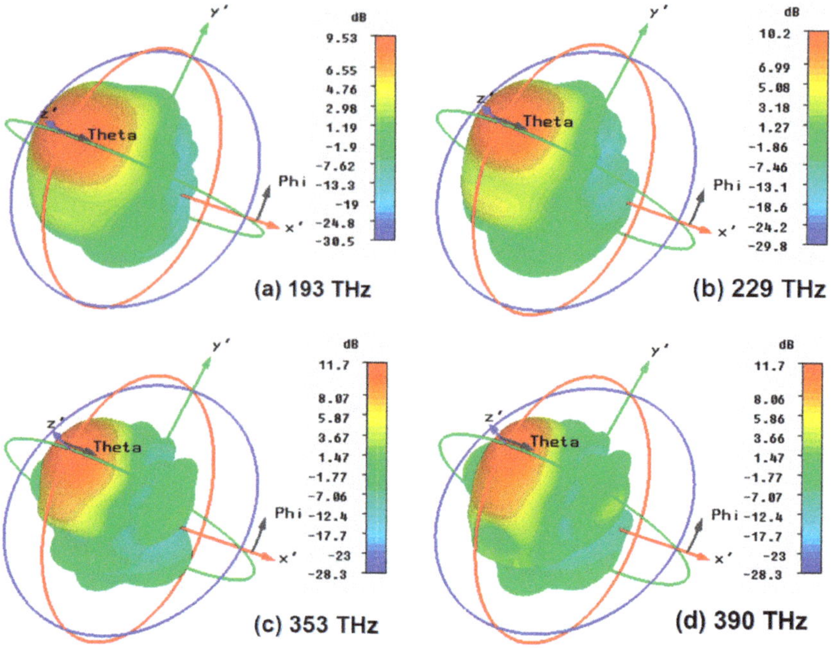

Figure 17. The 2 × 1 array 3D far field gain pattern at (**a**) 193 THz; (**b**) 229 THz; (**c**) 353 THz; (**d**) 390 THz.

Researchers have designed various antennas for nano-photonics applications. The proposed antenna has a higher gain and directivity to fully fill the same application compared with the state-of-the-art published work in Table 2.

Table 2. Comparison of the proposed antenna with the existing antennas.

References	[47]	[44]	[49]	[50]	[51]	[52]	This Work
Frequency (THz)	193.5/229/352.9	193.5	193.5/229/352.9	193.5	193.5	193.5	193.5/229/352.9/393
Gain (dBi)	5.03/4.64/2.02	5.6	4.67/4.8/7.26	7.99	7.5	9	8.02/8.41/10.6/11.4
Efficiency (%)	***	87	***	84.56	***	89	60
Bandwidth (GHz)	234	15.6	240	203	20	40	400

*** Not Given.

As the simulation analysis shows an increase in gain and directivity with the increase in the number of antenna elements, the higher-order array size is required to achieve a gain above 20 dBi. By performing the calculations using the array tool kit in CST MWS, which uses the array factor formula [53] to calculate the array pattern from a single nano-antenna, a 64 × 1 array is required to achieve a 3 dB bandwidth gain of 20 dBi. The polar plot and 3D radiation pattern of the 64 × 1 element array are depicted in Figures 18 and 19, which show that the beam width becomes narrower and a peak gain of about 26 dBi can be achieved. Table 3 presents a summary of the results with the single, 2 × 1, and 64 × 1 array nano-antennas for comparison.

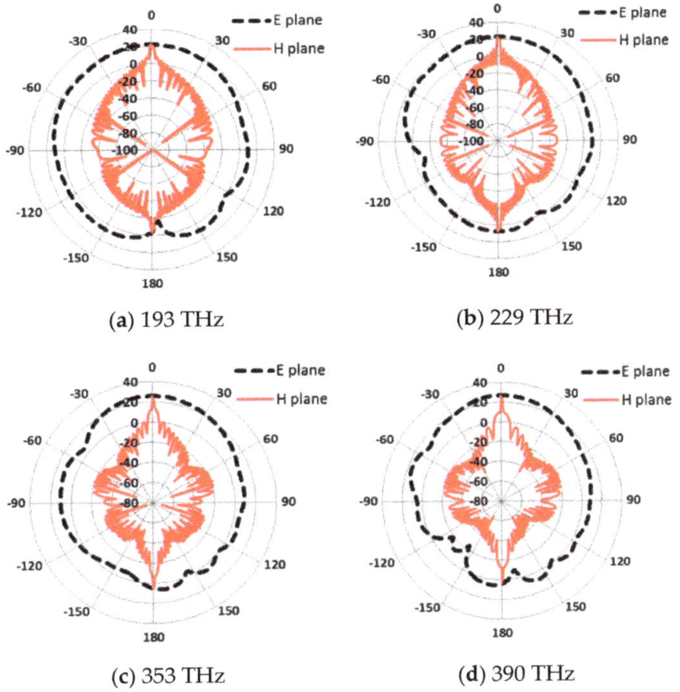

Figure 18. The 64 × 1 array gain plots in the E and H planes at (**a**) 193 THz; (**b**) 229 THz; (**c**) 353 THz; (**d**) 390 THz.

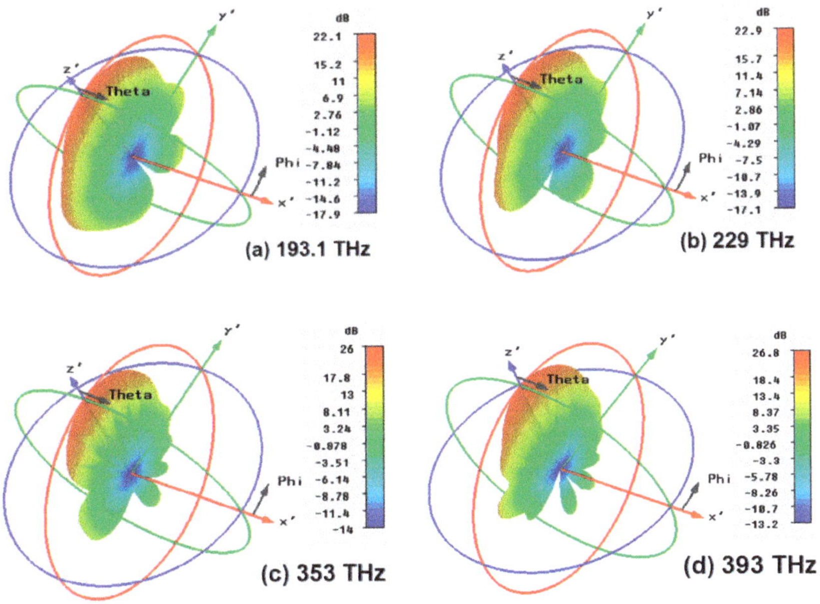

Figure 19. The 64 × 1 array 3D far field gain pattern at (**a**) 193 THz; (**b**) 229 THz; (**c**) 353 THz; (**d**) 390 THz.

Table 3. Summary of the overall results of gain and directivity.

Parameter	Gain (dBi)		
Frequency (THz)	**Single Element**	**2 × 1 Array**	**64 × 1 Array**
193	8.02/9.65	9.53/11.2	22.1/23.7
229	8.41/10.7	10.2/11.9	22.9/25.1
353	10.6/12.7	11.7/13.8	26/28.1
390	11.4/13.5	11.7/13.8	26.8/28.8

5. Conclusions

In this paper, a nature-inspired maple-leaf shaped optical nano-antenna is presented and fed with a hybrid plasmonic waveguide (HPW) to achieve a high gain with a broadband operation. The optimization of the proposed design provides a high gain of 11.8 dBi at THz frequencies and a bandwidth of 400 GHz. Moreover, the antenna is explored to operate in an array structure (2 × 1 and 64 × 1), in order to be used for optical energy harvesting applications with a satisfactory performance and gain above 26 dBi. The proposed design shows the ability to operate over a broad range of optical frequencies as shown by the simulation results, by covering all of the key optical communication band ranges from 660 to 6000 nm. Therefore, this design can be integrated with applications such as optical wireless communications in inter- and intra-chip devices and sensing.

Author Contributions: Conceptualization, I.A., S.U. (Shakir Ullah), and U.H.; data curation, J.u.d.; formal analysis, I.A., W.U. and S.U. (Sadiq Ullah); funding acquisition, S.K. and J.A.; methodology, I.A., S.U. (Shakir Ullah), J.u.d. and U.H.; project administration, S.K.; resources, J.A. and S.U. (Sadiq Ullah); software, S.U. (Shakir Ullah); supervision, S.U. (Sadiq Ullah); validation, W.U.; writing—original draft, U.H.; writing—review and editing, S.K. and J.A. All authors have read and agreed to the published version of the manuscript.

Funding: This work has been funded by King Saud University, Riyadh, Saudi Arabia through Researchers Supporting Project number (RSP-2021/58).

Institutional Review Board Statement: Not applicable.

Informed Consent Statement: Not applicable.

Data Availability Statement: Not applicable.

Acknowledgments: The authors sincerely appreciate funding from Researchers Supporting Project number (RSP-2021/58), King Saud University, Riyadh, Saudi Arabia.

Conflicts of Interest: The authors declare no conflict of interest.

References

1. Ullah, S.; Ahmad, I.; Raheem, Y.; Ullah, S.; Ahmad, T.; Habib, U. Hexagonal shaped CPW feed based frequency reconfigurable antenna for WLAN and sub-6 GHz 5G applications. In Proceedings of the IEEE 2020 International Conference on Emerging Trends in Smart Technologies (ICETST), Karachi, Pakistan, 26–27 March 2020; pp. 1–4.
2. Panda, R.A.; Kumari, P.; Naik, J.; Negi, P.; Mishra, D. Flower Shaped Patch with Circular Defective Ground Structure for 15 GHz Application. In *International Conference on Innovations in Bio-Inspired Computing and Applications*; Springer: Cham, Switzerland, 2019; pp. 243–249.
3. Abolade, J.O.; Konditi, D.B.O.; Dharmadhikary, V.M. Bio-inspired wideband antenna for wireless applications based on perturbation technique. *Heliyon* **2020**, *6*, e04282. [CrossRef]
4. De Oliveira, M.A.; da Costa, A.P.; Forte, G.G.S.; de Melo, P.-K.P.; Fontgalland, G.; Silva, P.-H.F.; Fontgalland, I.L. Using polar transformation to design a dissimilar antenna array inspired on four-leaf clover. In Proceedings of the 2018 IEEE Radio and Wireless Symposium (RWS), Anaheim, CA, USA, 15–18 January 2018; pp. 228–230.
5. Abolade, J.O.; Konditi, D.B.; Dharmadhikary, V.M. Compact Vitis vinifera-Inspired Ultrawideband Antenna for High-Speed Communications. *Int. J. Antennas Propag.* **2021**, *2021*, 9975884. [CrossRef]
6. da Silva Júnior, P.F.; Carlos, R.; Freire, S.; René Serres, A.J.; da Fonseca Silva, P.H.; Costa Silva, J. Bio-Inspired Antenna for UWB Systems. In Proceedings of the 2016 1st International Symposium on Instrumentation Systems, Circuits and Transducers (INSCIT), Belo Horizonte, Brazil, 29 August 2016–3 September 2016; p. 16408466.

7. Mesquita, M.D.S.; D'Assunção, A.G.; Oliveira, J.B.L.; Batista, Y.M.V. A New Conductive Ink for Microstrip Antenna and Bioinspired FSS Designs on Glass and Fiberglass Substrates. *J. Microw. Optoelectron. Electromagn. Appl.* **2019**, *18*, 227–245. [CrossRef]
8. Malik, R.; Singh, P.; Ali, H.; Goel, T. A Star Shaped Superwide Band Fractal Antenna for 5G Applications. In Proceedings of the 2018 3rd International Conference for Convergence in Technology (I2CT), Pune, India, 6–8 April 2018; pp. 1–6.
9. Anguera, J.; Puente, C.; Borja, C.; Soler, J. Fractal-Shaped Antennas: A Review. In *Wiley Encyclopedia of RF and MicrowaveEngineering*; Chang, K., Ed.; JohnWiley & Sons, Inc.: New York, NY, USA, 2005; Volume 2, pp. 1620–1635.
10. Melchiorre, L.; Marasco, I.; Niro, G.; Basile, V.; Marrocco, V.; D'Orazio, A.; Grande, M. Bio-Inspired Dielectric Resonator Antenna for Wideband Sub-6 GHz Range. *Appl. Sci.* **2020**, *10*, 8826. [CrossRef]
11. Marrocco, V.; Basile, V.; Marasco, I.; Niro, G.; Melchiorre, L.; D'Orazio, A.; Grande, M.; Fassi, I. Rapid Prototyping of Bio-Inspired Dielectric Resonator Antennas for Sub-6 GHz Applications. *Micromachines* **2021**, *12*, 1046. [CrossRef]
12. Bharadwaj, P.; Deutsch, B.; Novotny, L. Optical antennas. *Adv. Opt. Photonics* **2009**, *1*, 438–483. [CrossRef]
13. Duman, H. *Design and Fabrication of Resonant Nanoantennas on Chalcogenide Glasses for Nonlinear Photonic Applications*; Bilkent University: Ankara, Turkey, 2013.
14. Yuming, W. Analysis and Design of Nanoantennas. Ph.D. Dissertation, National University of Singapore, Singapore, 2010.
15. Haroyan, H.S.; Tadevosyan, V.R. High Gain Broadband Plasmonic Slot Nano-Antenna. *Int. J. Phys. Math. Sci.* **2015**, *9*, 1104–1107.
16. Sethi, W.T.; Vettikalladi, H.; Fathallah, H.; Himdi, M. Hexagonal dielectric loaded nantenna for optical ITU-T C-band communication. In Proceedings of the 2015 IEEE 11th International Conference on Wireless and Mobile Computing, Networking and Communications (WiMob), Abu Dhabi, United Arab Emirates, 19–21 October 2015; pp. 604–607.
17. Zhou, R.; Ding, J.; Arigong, B.; Lin, Y.; Zhang, H. Design of a new broadband monopole optical nano-antenna. *J. Appl. Phys.* **2013**, *114*, 184305. [CrossRef]
18. Kausar, A.S.M.Z.; Reza, A.W.; Latef, T.A.; Ullah, M.H.; Karim, M.E. Optical nano antennas: State of the art, scope and challenges as a biosensor along with human exposure to nano-toxicology. *Sensors* **2015**, *15*, 8787–8831. [CrossRef]
19. Rosner, B.; Peck, J.; van der Weide, D. Near-field antennas integrated with scanning probes for THz to visible microscopy: Scale modeling and limitations on performance. *IEEE Trans. Antennas Propag.* **2002**, *50*, 670–675. [CrossRef]
20. Xu, T.; Wu, Y.-K.; Luo, X.; Guo, L.J. Plasmonic nanoresonators for high-resolution colour filtering and spectral imaging. *Nat. Commun.* **2010**, *1*, 59. [CrossRef] [PubMed]
21. Tang, L.; Kocabas, S.E.; Latif, S.; Okyay, A.K.; Ly-Gagnon, D.S.; Saraswat, K.C.; Miller, D.A. Nanometre-scale germanium photodetector enhanced by a near-infrared dipole antenna. *Nat. Photonics* **2008**, *2*, 226–229. [CrossRef]
22. Rabienejhad, M.J.; Mazaheri, A.; Davoudi-Darareh, M. Design and optimization of nano-antenna for thermal ablation of liver cancer cells. *Chin. Phys. B* **2021**, *30*, 048401. [CrossRef]
23. Habib, U.; Aighobahi, A.; Wang, C.; Gomes, N.J. Radio over fiber transport of mm-Wave 2 × 2 MIMO for spatial diversity and multiplexing. In Proceedings of the IEEE International Topical Meeting on Microwave Photonics (MWP), Long Beach, CA, USA, 31 October 2016–3 November 2016; pp. 39–42.
24. Tapio, K.; Mostafa, A.; Kanehira, Y.; Suma, A.; Dutta, A.; Bald, I. A Versatile DNA Origami-Based Plasmonic Nanoantenna for Label-Free Single-Molecule Surface-Enhanced Raman Spectroscopy. *ACS Nano* **2021**, *15*, 7065–7077. [CrossRef] [PubMed]
25. Liu, Y.; Li, K.; Cao, S.; Xiong, G.; Zhu, L. Optimization Design of a Multi-slot Nanoantenna Based on Genetic Algorithm for Energy Harvesting. *Plasmonics* **2019**, *14*, 1577–1586. [CrossRef]
26. Mora-Ventura, B.; Sánchez, J.E.; González, G.; González, F.J. Thermal impedance analysis of nano-dipole linear arrays for energy harvesting applications. *Infrared Phys. Technol.* **2020**, *107*, 103332. [CrossRef]
27. Wu, Y.-M.; Li, L.-W.; Liu, B. Gold bow-tie shaped aperture Nanoantenna: Wide band near-field resonance and far-field radiation. *IEEE Trans. Magn.* **2010**, *46*, 1918–1921. [CrossRef]
28. Miroshnichenko, A.E.; Maksymov, I.S.; Davoyan, A.R.; Simovski, C.; Belov, P.; Kivshar, Y.S. An arrayed nanoantenna for broadband light emission and detection. *Phys. Status Solidi (RRL)-Rapid Res. Lett.* **2011**, *5*, 347–349. [CrossRef]
29. Biagioni, P.; Huang, J.S.; Duò, L.; Finazzi, M.; Hecht, B. Cross Resonant Optical Antenna. *Phys. Rev. Lett.* **2009**, *102*, 256801. [CrossRef] [PubMed]
30. Ahmad, I.; Ullah, S.; Ullah, S.; Habib, U.; Ahmad, S.; Ghaffar, A.; Alibakhshikenari, M.; Khan, S.; Limiti, E. Design and Analysis of a Photonic Crystal Based Planar Antenna for THz Applications. *Electronics* **2021**, *10*, 1941. [CrossRef]
31. Zainud-Deen, S.H.; Malhat, H.A.; El-Refaay, E.A. Polarization-Independent Reconfigurable Graphene Gas Sensor Using Crescent Plasmonic Antenna. *Plasmonics* **2020**, *15*, 1115–1122. [CrossRef]
32. Dong, Z.; Sun, C.; Si, J.; Deng, X. A tunable plasmonic nano-antenna based on metal–graphene double-nanorods. *Laser Phys. Lett.* **2018**, *15*, 056202. [CrossRef]
33. Yong, Z.; Gong, C.; Dong, Y.; Zhang, S.; He, S. Broadband localized electric field enhancement produced by a single-element plasmonic nanoantenna. *RSC Adv.* **2017**, *7*, 2074–2080. [CrossRef]
34. Iluz, Z.; Boag, A. Wideband dual Vivaldi nano-antenna with high radiation efficiency over the infrared frequency band. In Proceedings of the 2011 IEEE International Conference on Microwaves, Communications, Antennas and Electronics Systems (COMCAS), Tel Aviv, Israel, 5 July 2011; pp. 1–3.
35. Gramotnev, D.K.; Bozhevolnyi, S.I. Plasmonics beyond the diffraction limit. *Nat. Photonics* **2010**, *4*, 83–91. [CrossRef]
36. Economou, E.N. Surface plasmons in thin films. *Phys. Rev.* **1969**, *182*, 539. [CrossRef]

37. Maier, S.A.; Kik, P.G.; Atwater, H.A.; Meltzer, S.; Harel, E.; Koel, B.E.; Requicha, A.A.G. Local detection of electromagnetic energy transport below the diffraction limit in metal nanoparticle plasmon waveguides. *Nat. Mater.* **2003**, *2*, 229–232. [CrossRef]
38. Vernon, K.C.; Gramotnev, D.K.; Pile, D.F. Adiabatic nanofocusing of plasmons by a sharp metal wedge on a dielectric substrate. *J. Appl. Phys.* **2007**, *101*, 104312. [CrossRef]
39. Gielis, J. A generic geometric transformation that unifies a wide range of natural and abstract shapes. *Am. J. Bot.* **2003**, *90*, 333–338. [CrossRef] [PubMed]
40. Yousefi, L. Highly directive hybrid plasmonic leaky wave optical nano-antenna. *Prog. Electromagn. Res. Lett.* **2014**, *50*, 85–90. [CrossRef]
41. Alam, M.Z. *Hybrid Plasmonic Waveguides: Theory and Applications*; University of Toronto: Toronto, ON, Canada, 2012.
42. Ctyroký, J.; Kwiecien, P.; Richter, I. Analysis of hybrid dielectric-plasmonic slot waveguide structures with 3D Fourier Modal Methods. *J. Eur. Opt. Soc.-Rapid Publ.* **2013**, *8*, 13024. [CrossRef]
43. Yang, Y.; Li, Q.; Qiu, M. Broadband nanophotonic wireless links and networks using on-chip integrated plasmonic antennas. *Sci. Rep.* **2016**, *6*, 19490. [CrossRef]
44. Yousefi, L.; Foster, A.C. Waveguide-fed optical hybrid plasmonic patch nano-antenna. *Opt. Express* **2012**, *20*, 18326–18335. [CrossRef] [PubMed]
45. CST Microwave Studio CST of America, Inc. Available online: http://www.cst.com (accessed on 17 September 2021).
46. Haque, A.; Reza, A.W.; Kumar, N.; Ramiah, H. Slotting effect in designing circular edge bow-tie nano antenna for energy harvesting. In Proceedings of the 2015 IEEE Conference on Open Systems (ICOS), Melaka, Malaysia, 24–26 August 2015; pp. 46–50.
47. Saad-Bin-Alam, M.; Khalil, M.I.; Rahman, A.; Chowdhury, A.M. Hybrid Plasmonic Waveguide Fed Broadband Nanoantenna for Nano photonic Applications. *IEEE Photonics Technol. Lett.* **2015**, *27*, 1092–1095. [CrossRef]
48. Bozzetti, M.; de Candia, G.; Gallo, M.; Losito, O.; Mescia, L.; Prudenzano, F. Analysis and design of a solar rectenna. In Proceedings of the IEEE International Symposium on Industrial Electronics (ISIE 2010), Bari, Italy, 4–7 July 2010; pp. 2001–2004.
49. Nourmohammadi, A.; Nikoufard, M. Ultra-Wideband Photonic Hybrid Plasmonic Horn Nanoantenna with SOI Configuration. *Silicon* **2020**, *12*, 193–198. [CrossRef]
50. Sharma, P.; Vishwakarma, D.K. Long Range Multilayer Hybrid Plasmonic Waveguide Components and Integrated Circuit. *IEEE Trans. Nanotechnol.* **2019**, *18*, 940–947. [CrossRef]
51. Malheiros-Silveira, G.N.; Wiederhecker, G.S.; Hernández-Figueroa, H.E. Dielectric resonator antenna for applications in nanophotonics. *Opt. Express* **2013**, *21*, 1234–1239. [CrossRef]
52. Nikoufard, M.; Nourmohammadi, A.; Esmaeili, S. Hybrid Plasmonic Nanoantenna with the Capability of Monolithic Integration with Laser and Photodetector on InP Substrate. *IEEE Trans. Antennas Propag.* **2018**, *66*, 3–8. [CrossRef]
53. Balanis, C.A. *Antenna Theory: Analysis and Design*, 4th ed.; John Wiley & Sons: New York, NY, USA, 2016.

applied sciences

MDPI

Article

Burst Channel Error Reduction Based on Interleaving for Efficient High-Speed Wireless Communication

Fatma H. El-Fouly [1,*], Rabie A. Ramadan [2,3,*], Fathi E. Abd El-Samie [4,5,*], Mnaouer Kachout [2,6], Abdullah J. Alzahrani [2] and Jalawi Sulaiman Alshudukhi [2]

1 Department of Communication and Computer Engineering, Higher Institute of Engineering, El-Shorouk Academy, El-Shorouk City 11937, Egypt
2 College of Computer Science and Engineering, University of Hail, Hail 50141, Saudi Arabia; m.kachout@uoh.edu.sa (M.K.); aj.alzahrani@uoh.edu.sa (A.J.A.); j.alshudukhi@uoh.edu.sa (J.S.A.)
3 Department of Computer Engineering, Faculty of Engineering, Cairo University, Cairo 12613, Egypt
4 Department of Electronics and Electrical Communications, Faculty of Electronic Engineering, Menoufia University, Menouf 32952, Egypt
5 Department of Information Technology, College of Computer and Information Sciences, Princess Nourah Bint Abdulrahman University, Riyadh 11564, Saudi Arabia
6 Innov'COM, Sup'Comp, Carthage University, Tunis 1054, Tunisia
* Correspondence: fatma_elfoly@yahoo.com (F.H.E.-F.); rabie@rabieramadan.org (R.A.R.); fathi_sayed@yahoo.com (F.E.A.E.-S.)

Abstract: Recently, the demand for reliable and high-speed wireless communication has rapidly increased. Orthogonal frequency division multiplexing (OFDM) is a modulation scheme that is the newest competitor against other modulation schemes used for this purpose. OFDM is mostly used for wireless data transfer, although it may also be used for cable and fiber optic connections. However, in many applications, OFDM suffers from burst errors and high bit error rates. This paper presents the utilization of a helical interleaver with OFDM systems to efficiently handle burst channel errors and allow for Bit Error Rate (BER) reduction. The paper also presents a new interleaver, FRF, the initial letters of the authors' names, for the same purpose. This newly proposed interleaver summarizes our previous experience with many recent interleavers. Fast Fourier transform OFDM (FFT-OFDM) and Discrete Wavelet Transform OFDM (DWT-OFDM) systems are used to test the efficiency of the suggested scheme in terms of burst channel error removal and BER reduction. Finally, the general complexity of the FRF interleaver is different from that of the helical interleaver in terms of hardware requirements. The performance of the proposed scheme was studied over different channel models. The obtained simulation results show a noticeable performance improvement over the conventional FFT-OFDM and the FFT-OFDM systems with the helical interleaver. Finally, the disadvantage of the proposed FRF interleaver is that it is more complex than the helical interleaver.

Keywords: OFDM; FRF interleaver; helical interleaver; deterministic interleaver; FFT; DWT

Citation: El-Fouly, F.H.; Ramadan, R.A.; Abd El-Samie, F.E.; Kachout, M.; Alzahrani, A.J.; Alshudukhi, J.S. Burst Channel Error Reduction Based on Interleaving for Efficient High-Speed Wireless Communication. *Appl. Sci.* **2022**, *12*, 3500. https://doi.org/10.3390/app12073500

Academic Editor: Jaume Anguera

Received: 10 January 2022
Accepted: 21 March 2022
Published: 30 March 2022

1. Introduction

Wireless communication systems are extremely necessary to support high quality of service and high data rates. Channel frequency selectivity, multipath fading, and inter-symbol interference (ISI) often impair wireless channel communications. This substantially degrades both service quality and data rates [1]. OFDM is a multicarrier modulation technology that processes data using multiple orthogonal sub-carriers from the same source. The great spectral efficiency of orthogonality-based OFDM systems, their resistance to frequency-selective fading, and their easy equalizer implementation have recently sparked considerable interest. It has also become the widely recognized modulation scheme for high-data-rate communication over wireless connections [2]. The multiplexing is performed on the transmitter and receiver signals using inverse fast Fourier transform (IFFT) and fast Fourier transform (FFT) in the standard OFDM system. In addition, recent research on multicarrier modulation performance has explored the use of wavelet-based OFDM as an appealing alternative to the Fourier-based OFDM system [3–8]. Because there is no cyclic prefix requirement, the wavelet-based OFDM system offers better spectral efficiency, has very tight side lobes, and shows enhanced BER performance [9–12]. Another topic that will be considered in this work is the performance enhancement of data transmission in OFDM systems through data interleaving. It is known that data transmission over wireless channels may suffer from severe adverse conditions, especially burst errors, where errors are likely to occur in clusters. Interleavers have been used extensively in digital communication systems to disperse these burst errors [13,14]. Some simple interleaving techniques have been proposed [15–17]. The performance of such simple interleavers is limited. Therefore, there is a need for more powerful interleavers.

This work depends on using helical interleavers with OFDM systems to effectively manage burst channel errors and reduce BER. In addition, the paper presents a new interleaver, FRF, for the same purpose. This suggested interleaver summarizes our prior experience with several recent interleavers. There are two types of systems that are used to test the proposed interleaver: FFT-OFDM and DWT-OFDM. The FRF interleaver is more complicated than the helical interleaver in terms of hardware. It can also combat the channel effects with neither complicated coding schemes for error detection and correction nor adaptation of the modulation scheme. Another advantage of the proposed interleaver is that it achieves a degree of encryption in the transmitted data, which adds more security to the data transmission process.

Therefore, the contribution of this work could be summarized as follows:

1. Introducing a new interleaver that manages burst channel errors and reduces BER. Such an interleaver can be used in critical systems where power consumption is essential.
2. The new interleaver combats the channel effects with neither complicated coding schemes for error detection and correction nor adaptation of the modulation scheme.
3. The new interleaver achieves a degree of encryption in the transmitted data.

The paper is structured as follows. Section 2 introduces the OFDM and interleaver overview. Then, Section 3 describes the proposed interleaver. In addition, the proposed system model is described in Section 4. The simulation results are given in Section 5. Finally, the conclusions are presented in Section 6.

2. Burst Error, OFDM, and Interleaving

Adding redundancy to information sequences increases message delivery success rates in communication systems. Burst errors occur in a continuous portion of the received data. Burst errors represent a typical kind of interference in radio communications. For example, let u be an information sequence.

u = ABC

Let the error correction coding (ECC) duplicate each symbol as follows:

$v = A_1A_2A_3B_1B_2B_3C_1C_2C_3$

These data are now wirelessly sent. For a three-symbol-length burst of interference during transmission, this yields an error stream (X):

$R = A_1A_2A_3XXXC_1C_2C_3$

Despite the redundancies, symbol B is now irrecoverable due to the loss of all copies. If the error exceeds the repeated symbols, the error-correcting algorithm fails to retrieve the original data sequence. Let the redundant symbols be scrambled randomly within the stream, for example:

$V = A_2B_3C_2A_1B_1C_3A_2B_2C_1$

The same burst interference that results in the sequence can be as follows:

$R = A_2B_3C_2XXXA_2B_2C_1$

As a result, we obtain some duplicates of all the symbols, from which the original symbols and information sequence may be reconstructed. Therefore, interleavers are responsible for the scrambling of the message information. The detailed concept of interleavers will be stated in later subsections.

2.1. FFT-OFDM

As can be seen in Figure 1, depicting the OFDM transmitter, the data generator produces $\{d_k\}$ in a random binary form. Firstly, constellation mapping is implemented. The quadrature phase-shift keying (QPSK) modulator is used for this work to map the data to appropriate QPSK symbols X_m. The serial-to-parallel converter then transforms the high-speed information symbols into N sub-channel parallel information. The parallel data symbols are fed into IFFT to generate the multicarrier OFDM signal as follows [18]:

$$x_k = \frac{1}{\sqrt{N}} \sum_{m=0}^{N-1} x_m e^{j2\pi km/N},\ 0 \leq m \leq N-1 \quad (1)$$

where $\{x_k \mid 0 \leq k \leq N-1\}$ is a sequence in the discrete time domain, and $\{X_m \mid 0 \leq m \leq N-1\}$ are complex numbers in the discrete frequency domain.

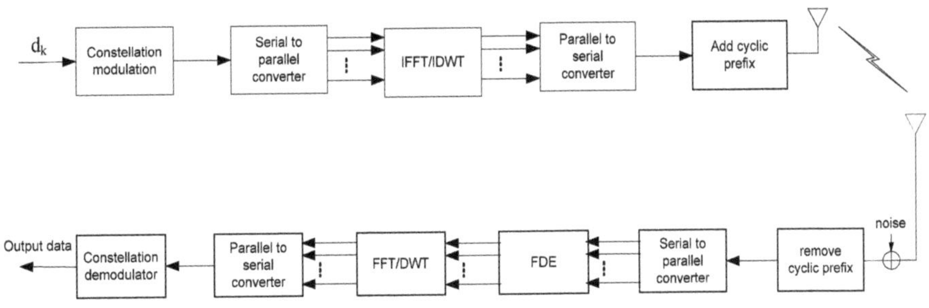

Figure 1. OFDM transceiver [18].

To prevent the ISI that happens in multipath channels, guard intervals are placed between frames. These intervals are either added zeros or a cyclic prefix (CP). The most commonly used strategy is the addition of a CP [19,20]. The CP is a copy of the IFFT last N samples at the start of the OFDM frame. The multipath versions of a symbol do not interfere with the next symbol by adding the CP to each OFDM symbol with a length longer than the channel length. Consequently, the impact of ISI is totally abolished, and at the same time, the channel appears to be circular, allowing a circular convolution to be implemented between the OFDM signal and the channel impulse response. This method is reversed at the receiver side to get the decoded information. The CP is detached, and the frequency domain equalizer (FDE) is used. The FDE relies on a CP guard interval between successive data blocks. The FDE can invert the channel completely. The benefit of FDE is that the complexity is comparatively low [21,22].

Equalizer coefficients are calculated by Minimum Mean Square Error (MMSE) or Zero Force (ZF) technique [23]. The MMSE equalizer requires the Signal-to-Noise-Ratio (SNR) estimation, which allows a trade-off between channel inversion and noise enhancement [24]. The FFT is applied at the receiver to reconstruct the signal as follows [18]:

$$X_m = \sum_{k=0}^{N-1} x_k e^{-j2\pi km/N},\ 0 \leq k \leq N-1 \quad (2)$$

2.2. DWT-OFDM

The inverse discrete wavelet transformation (IDWT) and the DWT take the place of the IFFT and FFT, respectively, in the DWT-OFDM system. The output of the IDWT can be represented as [25]:

$$s(k) = \sum_{m=0}^{\infty} \sum_{n=0}^{\infty} S_m^n 2^{m/2} \psi(2_k^m - n) \quad (3)$$

where S_m^n are the wavelet coefficients and $\psi(t)$ is the wavelet function with compression factor m and shift n for each sub-carrier. The process is reversed at the receiver side. The output of the DWT could be formed as shown in Equation (4).

$$S_m^n = \sum_{k=0}^{N-1} s(k)2^{m/2}\psi(2_k^m - n) \quad (4)$$

In the wavelet-based OFDM system, the MMSE equalizer achieves a significantly lower BER compared to that of the zero-forcing (ZF) equalizer. Therefore, it is considered in our work.

2.3. Interleavers

Bit errors are more likely to occur in bursts on wireless channels because of the fading nature of the channels and the impulsive noise. The goal of interleaving is to disperse bursts of errors throughout the data stream. It rearranges symbols to be transmitted in a certain order. The receiver uses the reverse rule to revert the sequence [26,27]. This section focuses only on the interleavers related to our proposal in this paper.

2.3.1. Helical Interleaver

The helical interleaver algorithm can be summarized as follows [28–31]:

(i) Primary interleaver generation: This refers to arranging the data sequence in a matrix with N_r rows and N_c columns, such that N_r. N_c = L. L, in this context, refers to the length of the primary interleaver, as given in Figure 2a.
(ii) Helical interleaver: It is constructed based on the primary interleaver through reading the interleaver indices column-wise, as given in Figure 2b.
(iii) By cyclically reading the interleaver indices from the diagonals of a matrix with decreasing slope, other interleavers can be generated, as indicated in Figure 2c.

The *ith* helical interleaver can be represented in Equation (5) as follows:
The *i'th* helical interleaver can be represented in Equation (5) as follows:

$$\pi[k] = \pi[l_{modL}],\ 0 \le k < L \quad (5)$$

where, $l = k_{modN_r}.N_c + \left(\left[\frac{k}{N_r}\right] + \left(k_{modN_r}.(i-1)\right)\right)$.

In Equation (5), it is seen that the helical interleaver can be managed in a very short time. If the parameters are chosen correctly, the interleaver indices can be spread out quite well. Now, the best way to even better optimize Equation (5) is to add layer-specific shifts to it, as given in Equation (6).

$$\pi[k] = \pi\left[(l + i.S)_{modN_c}\right] \quad (6)$$

where S is a constant integer, which describes the shift between the interleavers.

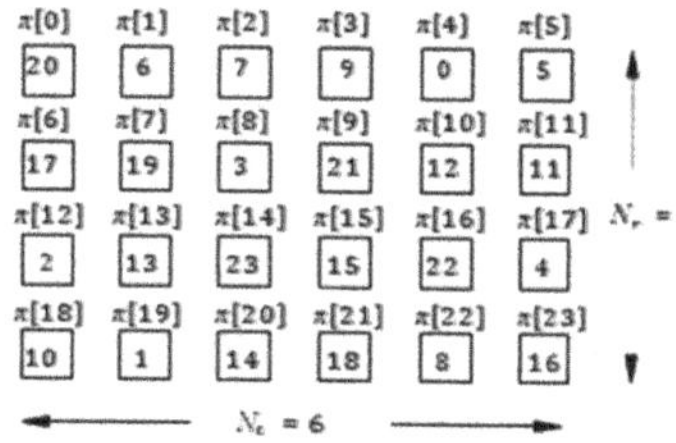

(a)

Figure 2. *Cont.*

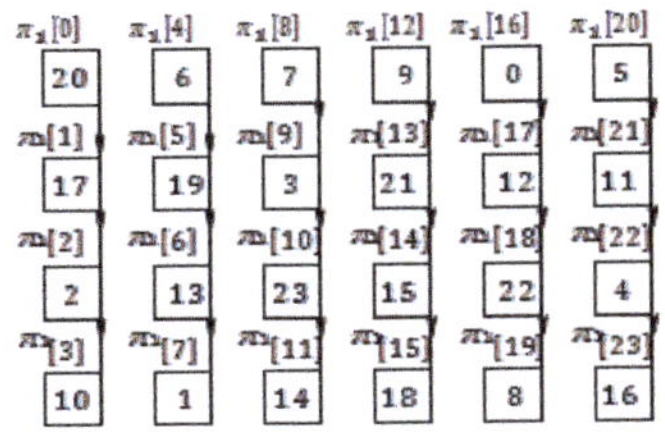

(b)

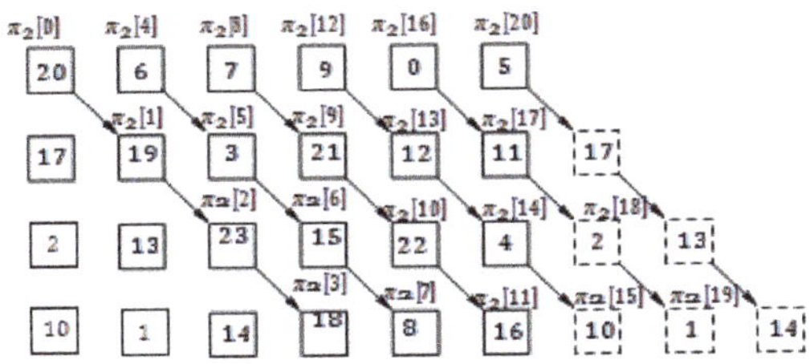

(c)

Figure 2. Generation of 4 × 6 helical interleavers. (**a**) Master interleaver written in matrix form, π = [20, 6, 7, ... , 8, 16]. (**b**) First helical interleaver written in matrix form, π_1 = [20, 17, 2, ... , 4, 16]. (**c**) Second helical interleaver written in matrix form, π_2 = [20, 19, 23, ... , 13, 14] [28,29].

2.3.2. Other Interleavers

There are many other interleavers based on the different standards summarized in Figure 3 [32]. In addition, Table 1 summarizes the interleaver algorithms.

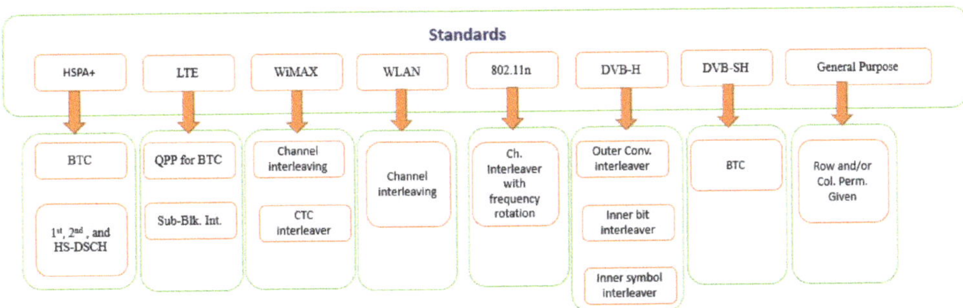

Figure 3. Other interleavers [32].

Table 1. Most of the current interleaver standards [32].

Interleaver Type	Algorithm
BTC	Multi-Step computation including intra-row permutation computation $S(j) = (v \times S(j-1))\ \%p;\ r(i) = T(q(i));$ $U(i,j) = S((j \times r(i)))\ \%(p-1);\ q\ mod(i) = r(i)\%(p-1);$ $RA(i,j) = \{RA(i,j-1) + q\ mod(i)\}\%(p-1);$ $I_{i,j} = \{C \times r(i)\} + U(i,j)$
1st, 2nd, and HS-DSCH	Standard block interleaving with different column permutations $\pi(k) = (p\lfloor \frac{k}{R} \rfloor + C \times (k\%R))\ \%K_\pi$
QPP for BTC	$I_{(x)} = (f_1 \cdot x + f_2 \cdot x^2)\ \%$
Sub-Blk. Int.	Standard block interleaving with given column permutations

Table 1. *Cont.*

Interleaver Type	Algorithm
Channel interleaving	$Two-$ step permutation $M_k = \left(\frac{N}{d}\right) \times (k\%d) + \left\lfloor \frac{k}{d} \right\rfloor$; and $J_k = s \times \left\lfloor \frac{M_k}{s} \right\rfloor \left(\left(M_k + N - \left\lfloor d \times \frac{M_k}{N} \right\rfloor \right) \%s \right)$
CTC interleaver	$I_{(x\%4=2)} = (P_0 x + 1 + P_1)\,\%N;\ I_{(x\%4=3)} = \left(P_0 x + 1 + \frac{N}{2} + P_3\right)\%N$
Ch. Interleaver with frequency rotation	Two step permutation as above, with extra frequency Interleaving i.e., $R_k = \left[J_k - \left\{ \left(((i_{ss}-1) \times 2)\,\%3 + 3\left\lfloor \frac{i_{ss}-1}{3} \right\rfloor \right) \times N_{ROT} \times N_{BPSC} \right\} \right]\%N$
Outer Conv. interleaver	Permutation defined by depth of first branch (M) and number of total branches
Inner bit interleaver	Six parallel interleavers with different cyclic shift $H_e(w) = (w + \Delta)\,\%126;\ where\ \Delta = 0,\ 63,\ 105,\ 42,\ 21,$ and 84
Inner symbol interleaver	$y_{H(q)} = x_q$ for even symbols; $y_q = x_{H(q)}$ for odd symbols; where $H(q) = (i\%2) \times 2^{N_r-1} + \sum_{j=0}^{N_r-2} R_i(j) \times 2^j$;
BTC	$R_c(j) = \{R_c(j-1) + Inc(j)\}\,\%32$; and $I(i,j) = \{T_{bas}(j) + M_1(i-1,j)\}\,\%C_T$

3. Proposed Interleavers

This section presents the proposed interleavers, including the two-dimensional prime interleaver and the newly proposed FRF interleaver.

3.1. TWO-Dimensional Deterministic Interleaver Design

The main idea behind this interleaver is expanding the 1-D deterministic interleaver into 2-D [33]. The proposed 2-D deterministic interleaver works as follows, assuming the case of N_r rows and N_c columns:

- First, interleaving is split into row-wise and column-wise cases.
- The seeds for row-wise and column-wise interleavers are assumed to be two prime numbers. After interleaving, bits will be located as follows:

$$\begin{array}{cc} row-wise & column-wise \\ 0 \rightarrow 0 & 0 \rightarrow 0 \\ 1 \rightarrow (1 \times p_{row}) mod\ N_r & 1 \rightarrow (1 \times p_{col}) mod\ N_c \\ 2 \rightarrow (2 \times p_{row}) mod\ N_r & 2 \rightarrow (2 \times p_{col}) mod\ N_c \\ \vdots & \vdots \\ N_r - 1 \rightarrow ((N_r - 1) * p_{row})\ mod\ N_r & N_c - 1 \rightarrow ((N_c - 1) * p_{col})\ mod\ N_c \end{array}$$

  where p_{row} and p_{col} are row-wise and column-wise seeds, respectively.
- Finally, the new locations are mapped back into the two-dimensional interleaver, resulting in two-dimensional interleaved bits.

For instance, consider an 8 × 8 2-D deterministic interleaver with p_{row} = 3 and p_{col} = 5. The new locations of the bits will be as follows:

$$\begin{array}{cc} row-wise & column-wise \\ 0 \rightarrow 0 & 0 \rightarrow 0 \\ 1 \rightarrow (1 \times 3) mod\ 8 = 3 & 1 \rightarrow (1 \times 5) mod\ 8 = 5 \\ 2 \rightarrow (2 \times 3) mod\ 8 = 6 & 2 \rightarrow (2 \times 5) mod\ 8 = 2 \\ 3 \rightarrow (3 \times 3) mod\ 8 = 1 & 3 \rightarrow (3 \times 5) mod\ 8 = 7 \\ 4 \rightarrow (4 \times 3) mod\ 8 = 4 & 4 \rightarrow (4 \times 5) mod\ 8 = 4 \\ 5 \rightarrow (5 \times 3) mod\ 8 = 7 & 5 \rightarrow (5 \times 5) mod\ 8 = 1 \\ 6 \rightarrow (6 \times 3) mod\ 8 = 2 & 6 \rightarrow (6 \times 5) mod\ 8 = 6 \\ 7 \rightarrow (7 \times 3) mod\ 8 = 5 & 7 \rightarrow (7 \times 5) mod\ 8 = 3 \end{array}$$

The arrangement of the bits prior to interleaving is seen in Figure 4a. By applying the new order of column- and row-wise interleaved bits acquired from the preceding calculation and mapping the locations, we get the bit arrangement shown in Figure 4b after interleaving.

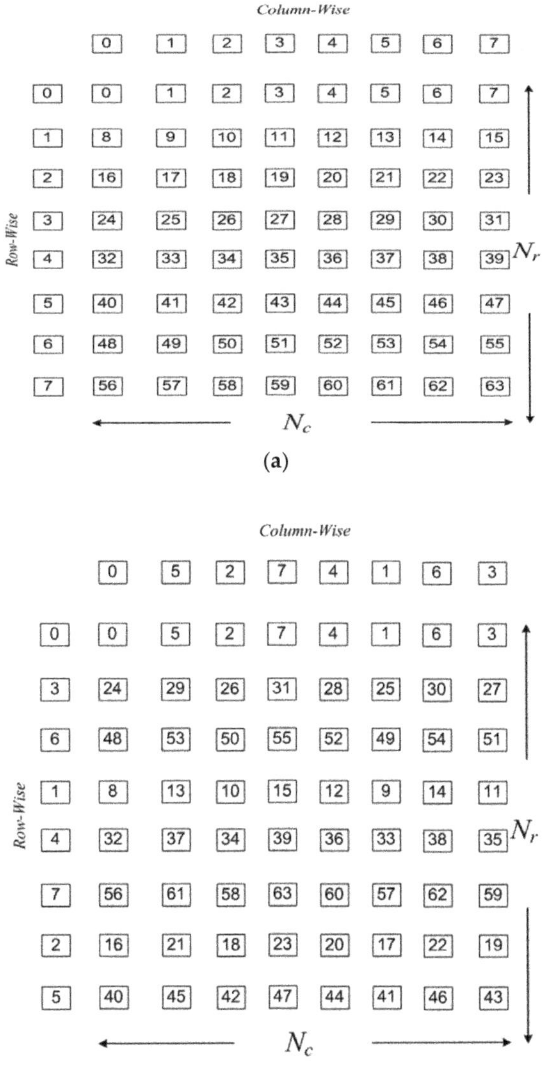

Figure 4. Arrangement of bits for an 8 × 8 channel (**a**) before interleaving and (**b**) after interleaving.

3.2. FRF Interleaver Design

Utilization of the proposed 2-D deterministic interleaver of size 8 × 8 with the helical interleaver produces highly strong randomization. In addition, based on our study, it has been found that the utilization of two-stage interleavers is not practical. Thus, this paper presents the novel FRF interleaver that can perform the task of two interleavers, but in a single stage.

The main concept of the proposed interleaver is stated as follows:

(i) First, we arrange the data in a matrix with N_r rows and N_c columns, such that $N_r \times N_c = L$, where L is the number of decoded bits, and each of them should be an integer and multiple of 8.

(ii) Using Equation (7), we obtain the new bit locations after interleaving:

$$\pi[k] = \pi[((ii + S \cdot jj) mod\ N_r) \cdot N_c + jj],\ 0 \leq k < L \tag{7}$$

where S is a constant integer, and

$$\begin{aligned} ii &= ((i \cdot 3) mod\ 8) + (i\ mod\ 8) + 1\ , i = 0, 1, \cdots, N_r - 1 \\ jj &= ((j \cdot 3) mod\ 8) + (j\ mod\ 8) + 1\ , j = 0, 1, \cdots, N_c - 1 \end{aligned}$$

Now, we take a look at how the helical and FRF interleaving mechanisms can correct error bursts. An example of the interleaving of an (8 × 8) square matrix for S = 2 is given in Figure 5. Assume a burst of errors affecting four consecutive samples (1-D error burst) as shown in Figure 5b,c with shades. After helical and FRF de-interleaving, the error burst is effectively spread among four different rows, resulting in a small effect for the 1-D error burst as shown in Figure 5d,e. With a single error correction capability, it is obvious that no decoding error will result from the presence of such a 1-D error burst. This simple example demonstrates the effectiveness of the helical and FRF interleaving mechanisms in combating 1-D bursts of errors. Let us examine the performance of the helical and FRF interleaving mechanisms when a 2-D (2 × 2) error burst occurs, as shown in Figure 5b,c with shades. Figure 5d,e show that although the two interleaving mechanisms effectively spread the 2 × 2 error burst, the FRF interleaver has a stronger randomization ability than the helical interleaver. As a result, a better BER performance can be achieved with the proposed FRF interleaving mechanism.

S_{1}	S_{2}	S_{3}	S_{4}	S_{5}	S_{6}	S_{7}	S_{8}
S_{9}	S_{10}	S_{11}	S_{12}	S_{13}	S_{14}	S_{15}	S_{16}
S_{17}	S_{18}	S_{19}	S_{20}	S_{21}	S_{22}	S_{23}	S_{24}
S_{25}	S_{26}	S_{27}	S_{28}	S_{29}	S_{30}	S_{31}	S_{32}
S_{33}	S_{34}	S_{35}	S_{36}	S_{37}	S_{38}	S_{39}	S_{40}
S_{41}	S_{42}	S_{43}	S_{44}	S_{45}	S_{46}	S_{47}	S_{48}
S_{49}	S_{50}	S_{51}	S_{52}	S_{53}	S_{54}	S_{55}	S_{56}
S_{57}	S_{58}	S_{59}	S_{60}	S_{61}	S_{62}	S_{63}	S_{64}

(a)

1-D error burst

S_{1}	S_{18}	S_{35}	S_{52}	S_{5}	S_{22}	S_{39}	S_{56}
S_{9}	S_{26}	S_{43}	S_{60}	S_{13}	S_{30}	S_{47}	S_{64}
S_{17}	S_{34}	S_{51}	S_{4}	S_{21}	S_{38}	S_{55}	S_{8}
S_{25}	S_{42}	S_{59}	S_{12}	S_{29}	S_{46}	S_{63}	S_{16}
S_{33}	S_{50}	S_{3}	S_{20}	S_{37}	S_{54}	S_{7}	S_{24}
S_{41}	S_{58}	S_{11}	S_{28}	S_{45}	S_{62}	S_{15}	S_{32}
S_{49}	S_{2}	S_{19}	S_{36}	S_{53}	S_{6}	S_{23}	S_{40}
S_{57}	S_{10}	S_{27}	S_{44}	S_{61}	S_{14}	S_{31}	S_{48}

2-D error burst

(b)

Figure 5. *Cont.*

1-D error burst

S_1	S_{54}	S_{35}	S_{24}	S_5	S_{50}	S_{39}	S_{20}
S_{25}	S_{14}	S_{59}	S_{48}	S_{29}	S_{10}	S_{63}	S_{44}
S_{49}	S_{38}	S_{19}	S_8	S_{53}	S_{34}	S_{23}	S_4
S_9	S_{62}	S_{43}	S_{32}	S_{13}	S_{58}	S_{47}	S_{28}
S_{33}	S_{22}	S_3	S_{56}	S_{37}	S_{18}	S_7	S_{52}
S_{57}	S_{46}	S_{27}	S_{16}	S_{61}	S_{42}	S_{31}	S_{12}
S_{17}	S_6	S_{51}	S_{40}	S_{21}	S_2	S_{55}	S_{36}
S_{41}	S_{30}	S_{11}	S_{64}	S_{45}	S_{26}	S_{15}	S_{60}

2-D error burst

(c)

S_1	S_2	S_3	S_4	S_5	S_6	S_7	S_8
S_9	S_{10}	S_{11}	S_{12}	S_{13}	S_{14}	S_{15}	S_{16}
S_{17}	S_{18}	S_{19}	S_{20}	S_{21}	S_{22}	S_{23}	S_{24}
S_{25}	S_{26}	S_{27}	S_{28}	S_{29}	S_{30}	S_{31}	S_{32}
S_{33}	S_{34}	S_{35}	S_{36}	S_{37}	S_{38}	S_{39}	S_{40}
S_{41}	S_{42}	S_{43}	S_{44}	S_{45}	S_{46}	S_{47}	S_{48}
S_{49}	S_{50}	S_{51}	S_{52}	S_{53}	S_{54}	S_{55}	S_{56}
S_{57}	S_{58}	S_{59}	S_{60}	S_{61}	S_{62}	S_{63}	S_{64}

(d)

S_1	S_2	S_3	S_4	S_5	S_6	S_7	S_8
S_9	S_{10}	S_{11}	S_{12}	S_{13}	S_{14}	S_{15}	S_{16}
S_{17}	S_{18}	S_{19}	S_{20}	S_{21}	S_{22}	S_{23}	S_{24}
S_{25}	S_{26}	S_{27}	S_{28}	S_{29}	S_{30}	S_{31}	S_{32}
S_{33}	S_{34}	S_{35}	S_{36}	S_{37}	S_{38}	S_{39}	S_{40}
S_{41}	S_{42}	S_{43}	S_{44}	S_{45}	S_{46}	S_{47}	S_{48}
S_{49}	S_{50}	S_{51}	S_{52}	S_{53}	S_{54}	S_{55}	S_{56}
S_{57}	S_{58}	S_{59}	S_{60}	S_{61}	S_{62}	S_{63}	S_{64}

(e)

Figure 5. Helical and FRF interleaving of an 8 × 8 matrix. (**a**) The 8 × 8 matrix. (**b**) Data with error bursts after helical interleaving. (**c**) Data with error bursts after FRF interleaving. (**d**) Effect of error bursts after helical de-interleaving. (**e**) Effect of error bursts after FRF de-interleaving.

4. Proposed System Model

The main idea of the proposed system is to use a combination of (FFT/DWT)-OFDM with the proposed FRF interleaving mechanism. The FRF interleaver can be a potential candidate for practical OFDM systems due to its low computational complexity and good BER performance. The block diagram of the proposed (FFT/DWT)-OFDM with helical interleaving is shown in Figure 6. The conventional OFDM block is modified by adding an interleaving stage. Both the in-phase and quadrature fields of the OFDM signal (the output of IFFT/IDWT) are interleaved. *S* is supposed to be known by the receiver.

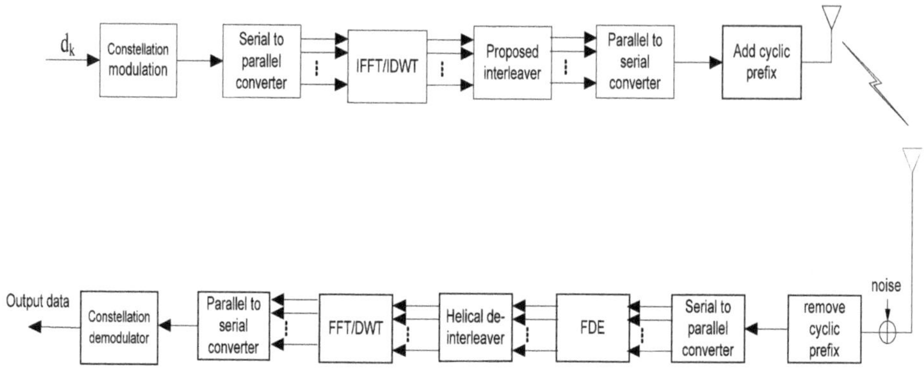

Figure 6. Block diagram of the proposed (FFT/DWT)-OFDM system model.

5. Simulation Results

The experiments in this section have been carried out using the MATLAB 7.5 program. Experiments have been conducted on an i5-2.3 GHz laptop running Microsoft Windows 7. In this section, computer simulations are presented to examine and evaluate the BER performances of different scenarios.

1. Different wavelet families are used in DWT- and FFT-OFDM.
2. Over the AWGN channel model, the proposed systems are compared to conventional (FFT/DWT)-OFDM systems.
3. With an exponential power delay profile Rayleigh fading channel without a Doppler effect, the proposed systems are compared to conventional (FFT/DWT)-OFDM systems.
4. In the presence of AWGN, the suggested systems are compared to conventional (FFT/DWT)-OFDM systems.
5. The suggested systems are compared to traditional (FFT/DWT)-OFDM systems over the AWGN outdoor channel.

To ensure the success of the proposed systems, we introduce the BER versus E_b/N_0 for all systems, where E_b is the energy per bit and N_0 is the noise power spectral density. The number of sub-carriers considered equals 512, with each sub-carrier having 16 symbols. The guard interval length is one-eighth of the symbol duration. QPSK or 4-QAM ($M = 4$) data symbols are used in the simulation experiments. In all later experiments except in the following subsection, we use the value of $S = 5$, where the evaluation results show that this value is the best for all experiments. Table 2 summarizes the simulation parameters.

Table 2. Simulation parameters.

Parameters	Values
Number of sub-carriers	512
Number of symbols with each sub-carrier	16
CP length	1/8 symbol duration
Modulation type	QPSK
Equalization	MMSE
Channel model	Rayleigh, SUI, and vehicular A outdoor channels

5.1. BER Performance Evaluation of DWT-OFDM and FFT-OFDM

An experiment was conducted in order to provide the wavelet with which the best performance in wireless communications can be obtained. In this experiment, the BER performances of (FFT/DWT)-OFDM systems have been evaluated with several mother wavelets such as Haar, Daubechies (db3), coiflets (coif1), symlets (sym3), biorthogonal (bior1.3), and reverse biorthogonal (rbio1.3) [34] over a multipath Rayleigh fading channel model in the absence of AWGN.

Figure 7 presents the BER vs. E_b/N_0. The figure shows that the DWT-OFDM outperforms the conventional OFDM for all chosen wavelet families. For instance, at BER = 2.4 × 10^{-4}, the DWT (Haar)-OFDM system provides E_b/N_0 gains of about 11 dB over the FFT-OFDM system. In addition, it can be seen that there is no significant difference between the kinds of wavelets except Haar, which performs slightly better than the other wavelets at high E_b/N_0 values. Therefore, the Haar mother wavelet will be the best choice for DWT-OFDM implementation. For instance, at BER = 1 × 10^{-7}, the Haar wavelet provides E_b/N_0 gains of about 1dB over coif1 and rbio1.3 and about 0.5 dB over bior1.3, db3, and sym3. After performing this experiment, we can recommend that the Haar wavelet gives better performance parameters for implementing DWT-OFDM.

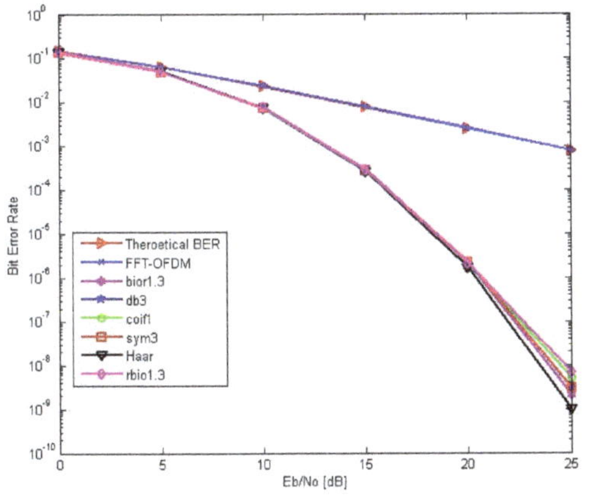

Figure 7. BER performance comparisons of the DWT-OFDM and the FFT-OFDM systems over Rayleigh channel model using several wavelets.

5.2. BER Performance Evaluation over AWGN Channel

This experiment evaluates the BER performances of the FFT-OFDM and the DWT (Haar)-OFDM with and without helical and FRF interleavers over the AWGN channel. Figure 8 shows the BER vs. E_b/N_0 for the helical (FFT/DWT)-OFDM and the FRF (FFT/DWT)-OFDM systems compared with the conventional (FFT/DWT)-OFDM over an AWGN channel. As shown in the results, there is no improvement, because there are no burst errors in the AWGN case. We can say that there is no need for the interleaver in this case.

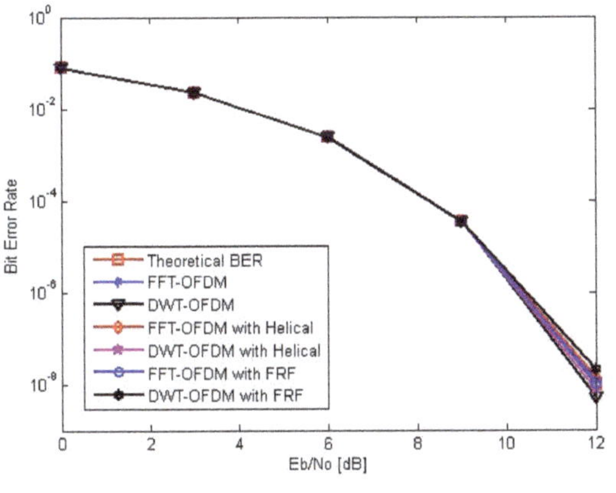

Figure 8. FFT−OFDM and DWT−OFDM BER performance over AWGN channel.

5.3. BER Evaluation over Rayleigh Fading Channel

In this experiment, the BER performances of the FFT-OFDM and the DWT (Haar)-OFDM with and without helical and FRF interleavers have been evaluated over a multipath Rayleigh fading channel with an exponential power delay profile and no Doppler effect in the presence of AWGN. Figure 9 shows the BER vs. E_b/N_0 for the helical (FFT/DWT)-OFDM and FRF (FFT/DWT)-OFDM systems compared with the conventional (FFT/DWT)-OFDM systems over a Rayleigh fading channel. It should be noted that the performance of the DWT-OFDM outperforms that of the conventional FFT-OFDM. This is justified as follows. The performance merit can be explained by the premise of the cyclic prefixing required in the FFT-OFDM system, which is not required in the DWT-OFDM modulation, since copying a certain portion of each of the transmit symbol lengths leads to some noise also being copied. This would lower the likelihood of decoding the transmitted bits correctly. Over the multipath channel, more orientation is imposed on the transmit signal by the channel impulse response, but the signals processed by the DWT scheme possess some sturdy flexibility in time and frequency coupled with the filtering mechanisms used in the wavelet transform. The wavelet filters decompose the signal into equal lengths of low-frequency band and high-frequency band, and likewise reconstruct them. Since the channel state is known to the receiver, the effect of the channel can be removed with some bearable error introduced by the system noise. Therefore, from Figure 9, we can see clearly that helical and FRF interleaving achieve the best results with DWT-OFDM and strong improvements with FFT-OFDM. The proposed FFT-OFDM system with FRF interleaving outperforms that with helical interleaving in terms of BER. The reason for this improvement can be explained as follows. As mentioned above, the FRF interleaver has better randomization capabilities than the helical interleaver; that is to say, the FRF interleaver generates permuted sequences with a lower correlation between their samples, which efficiently combats the channel effects without a need for complicated coding schemes for error detection.

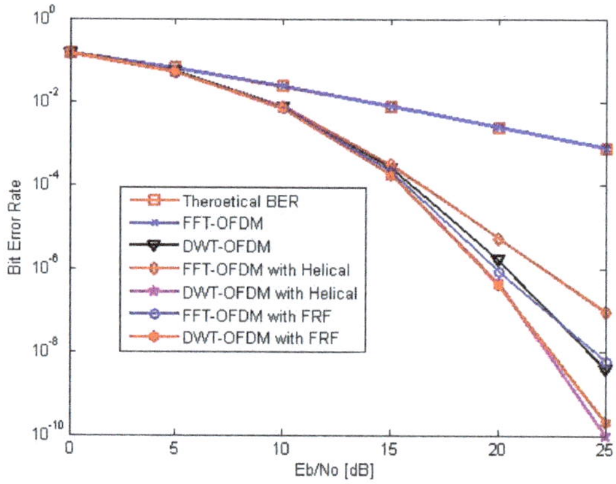

Figure 9. FFT−OFDM and DWT−OFDM BER performance over Rayleigh channel with exponential power delay profile channel without a Doppler effect.

On the other hand, the proposed DWT-OFDM system with the FRF interleaver provides approximately the same BER performance as that with the helical interleaver. In fact, the BER performance of communication systems can be improved due to the unique time–frequency localization feature of wavelets. As a result, there is no need to use interleavers with great randomization capabilities for the DWT-OFDM system. This is the reason why the FRF interleaver gives the same BER performance as that of the helical interleaver with the DWT-OFDM system. Thus, our recommendation is to use the proposed FRF interleaver for the FFT-OFDM system because of its ability to provide a good BER due to its inherent strong randomization ability. From the results shown in Figure 9, for example, at $E_b/N_0 = 20$ dB, the proposed FFT-OFDM system with the FRF interleaver achieves a

BER of about 8.8 × 10^{-7}. On the other hand, the BER value achieved in the FFT-OFDM system with the helical interleaver is 5.248 × 10^{-6}, while it is 2.557 × 10^{-3} in the case of the conventional FFT-OFDM system. This means that the proposed FRF interleaver can improve the BER performance compared to the BER achieved by the FFT-OFDM system with a helical interleaver and the conventional FFT-OFDM.

5.4. BER Performance Evaluation over Different Stanford University Interim (SUI) Channel Models

In this experiment, the BER performances of FFT-OFDM and DWT (Haar)-OFDM systems with and without helical and FRF interleavers are evaluated over different SUI channel models in the presence of AWGN. The SUI channel models considered are SUI-2, SUI-3, and SUI-6 [35]. The multipath profiles of the used SUI channels are summarized in Table 3.

Table 3. The multipath profiles of the used SUI channels.

Model	Delay / Gain	Tap1	Tap2	Tap3	Units
		L (No. of Taps) = 3			
SUI-2	Delay	0	0.4	1.1	μs
	Gain	0	−12	−15	dB
SUI-3	Delay	0	0.4	0.9	μs
	Gain	0	−5	−10	dB
SUI-6	Delay	0	14	20	μs
	Gain	0	−10	−14	dB

Figures 10–12 show the BER vs. E_b/N_0 for the helical (FFT/DWT)-OFDM and FRF (FFT/DWT)-OFDM systems compared with the conventional (FFT/DWT)-OFDM systems over the SUI-2, SUI-3, and SUI-6 channel models. As seen in Figures 10–12, the DWT-OFDM achieves a better performance than that of the FFT-OFDM for the same reasons described above. In addition, the proposed (FFT/DWT)-OFDM systems outperform the traditional (FFT/DWT)-OFDM systems. In addition, helical and FRF interleaving achieve the best results with the DWT-OFDM system. Therefore, we can say that the proposed systems have a high immunity to burst errors. The figures also show that the proposed FFT-OFDM system with FRF interleaving provides a better BER performance than that with the helical interleaving, but the proposed DWT-OFDM systems with FRF and helical interleavers provide approximately the same BER performance for the same reasons described above. This indicates that the FRF interleaver is more suitable for the FFT-OFDM system due to its strong randomization capability that leads to a good BER. From the previous results obtained from Figures 10–12, for example, at E_b/N_0 = 30 dB, the proposed FFT-OFDM system with an FRF interleaver achieves BERs of about 1.22 × 10^{-6}, 9.78 × 10^{-7}, and 1.25×10^{-6} for the SUI-2, SUI-3, and SUI-6 channel models, respectively. On the other hand, the BER values achieved by the FFT-OFDM system with the helical interleaver are 4.44 × 10^{-6}, 2.11 × 10^{-6}, and 4.67 × 10^{-6}, while they are 0.1333 × 10^{-3}, 0.2354 × 10^{-3}, and 0.2562 × 10^{-3} for the conventional FFT-OFDM system over the SUI-2, SUI-3, and SUI-6 channel models, respectively. This means that the proposed FRF interleaver can improve the BER performance compared to those of the FFT-OFDM system with the helical interleaver and the conventional FFT-OFDM.

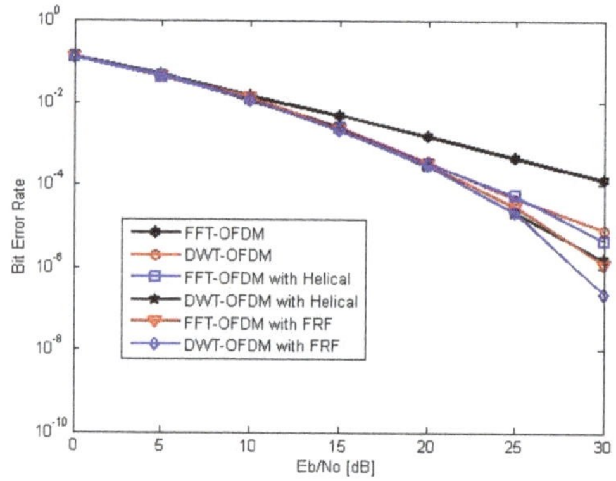

Figure 10. BER performance for the FFT−OFDM and DWT−OFDM systems over SUI-2 channel.

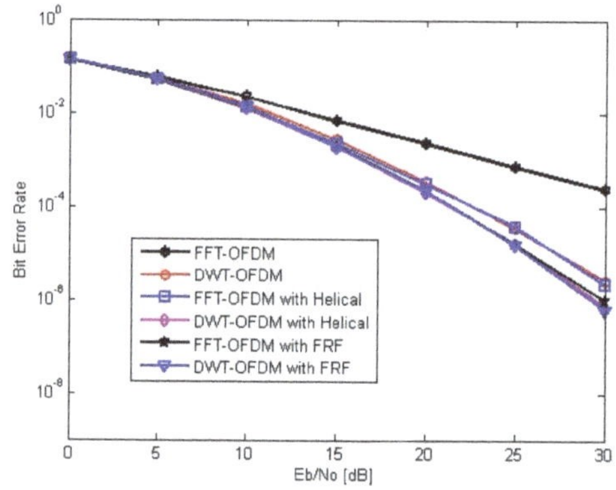

Figure 11. BER performance for the FFT−OFDM and DWT−OFDM systems over SUI-3 channel.

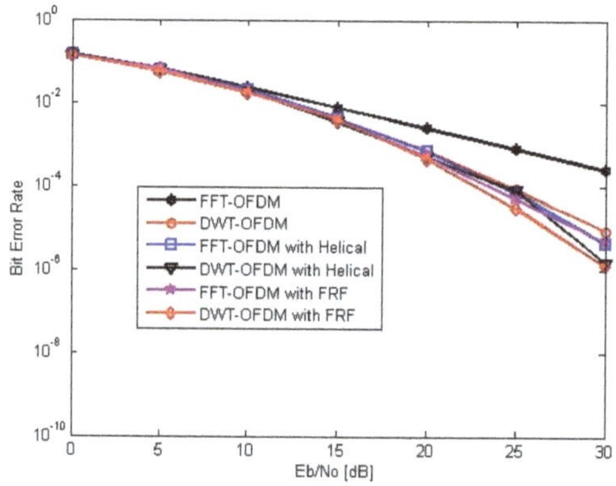

Figure 12. BER performance for the FFT−OFDM and DWT−OFDM systems over SUI-6 channel.

5.5. BER Performance Evaluation over Vehicular A Channel

In this experiment, the BER performances of FFT-OFDM and DWT (Haar)-OFDM with and without helical and FRF interleavers are evaluated over the vehicular A outdoor channel model in the presence of AWGN. The multipath profile of the vehicular A outdoor channel model is summarized in Table 4 [35].

Table 4. Multipath profile of the vehicular A outdoor channel model.

	Tap1	Tap2	Tap3	Tap4	Tap5	Tap6
Delay (ns)	0	310	710	1090	1730	2510
Power (dB)	0	−1	−9	−10	−15	−20

Figure 13 shows the BER vs. E_b/N_0 for the helical (FFT/DWT)-OFDM and FRF (FFT/DWT)-OFDM systems compared with the conventional (FFT/DWT)-OFDM over the vehicular A outdoor channel model. As can be seen in the figure, the proposed (FFT/DWT)-OFDM systems outperform the conventional (FFT/DWT)-OFDM systems. Hence, this confirms that the proposed systems have a high immunity to burst errors regardless of the channel models. From Figure 13, it is also clear that the proposed FFT-OFDM system with an FRF interleaver provides a better BER performance than that with the helical interleaver, but the proposed DWT-OFDM system with the FRF interleaver provides approximately the same BER performance as that with the helical interleaver for the same reasons mentioned above. Therefore, it is better to use the proposed FRF interleaver with the FFT-OFDM system for the same reasons described above. From the results obtained from Figure 13, at E_b/N_0 = 25 dB, the proposed FFT-OFDM system with an FRF interleaver achieves a BER of about 3.62×10^{-6}. On the other hand, the BER value achieved by the FFT-OFDM system with a helical interleaver is 1.335×10^{-5}, while it is 0.8369×10^{-3} in the case of the conventional FFT-OFDM system. This means that the proposed FRF interleaver can improve BER performance compared to those achieved by the FFT-OFDM system with the helical interleaver and the conventional FFT-OFDM system.

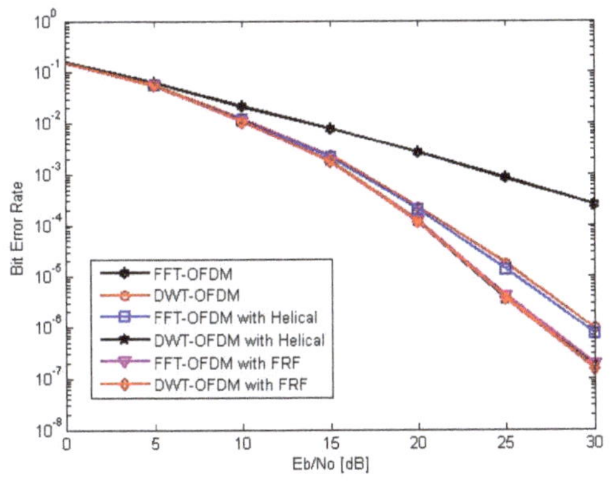

Figure 13. BER performance for FFT−OFDM and DWT−OFDM systems over vehicular A channel.

5.6. Overall Complexity Evaluation

In this set of experiments, the FRF interleaver is compared to the helical interleaver in terms of hardware complexity. The two interleavers were implemented in hardware using field-programmable gate arrays (FPGA) with a size of 128 × 128. Additionally, both interleavers were developed using Altera's Cyclone II FPGA board. Figures 14 and 15 indicate the overall complexity in terms of the total logic elements and memory bits. The FRF and helical interleavers' required logic elements were estimated using Altera's Quartus II tools. As indicated in the figures, the FRF interleaver needs more logic elements than the helical interleaver, but the total memory bits are almost the same for both of them. The FRF interleaver is, therefore, more complicated than the helical interleaver; that is to say, the FRF interleaver requires more energy in computations. However, in wireless networks such as wireless sensor networks, the energy consumption in communication is identified as the major source of energy consumption and costs significantly more than computation. Indeed, increasing the BER will increase the number of lost packets, leading to an increase in energy consumption due to packet retransmission that inevitably affects the network efficiency. Therefore, the FRF interleaver can improve the network throughput and energy efficiency compared with the others.

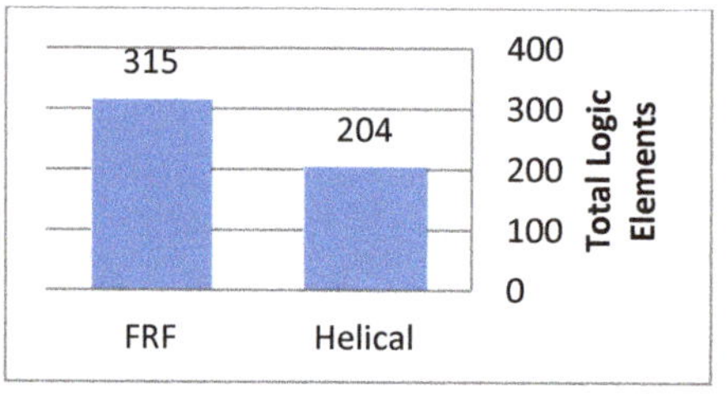

Figure 14. Total number of logic elements required for the FPGA implementation of the FRF and helical interleavers of size 128 × 128.

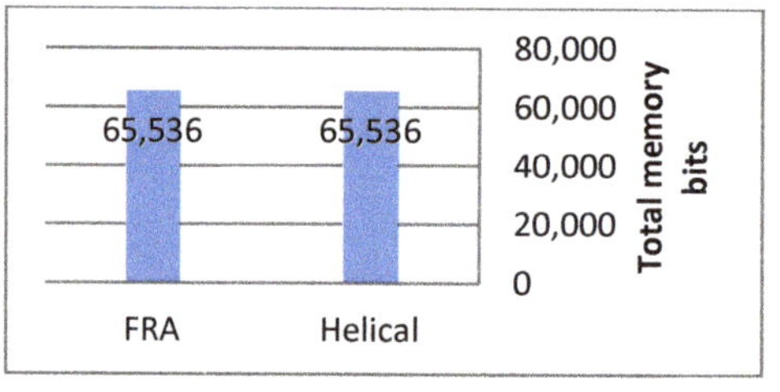

Figure 15. Total number of memory bits required for FPGA implementation of FRF and helical interleavers of size 128 × 128.

6. Conclusions

This research introduced an efficient helical interleaver and a novel FRF interleaver for OFDM systems to increase wireless communication reliability and data throughput. The proposed approach enhances the (FFT/DWT)-OFDM BER performance. The BER performance of the FFT-OFDM system was compared to that of the DWT-OFDM system for numerous wavelet families. The AWGN, multipath Rayleigh fading with exponential power delay spread and no Doppler influence, SUI-2, SUI-3, SUI-6, and vehicular A channel models have all been considered to assess the performance of the suggested approach. Finally, experimental findings indicate that higher-order *M*-QAM offers high data rates despite a weaker resilience to errors. The DWT-OFDM outperforms the traditional FFT-OFDM for all wavelet families. In addition, it is observed that the Haar wavelet is the best choice for DWT-OFDM implementation. Furthermore, according to the simulation results, it has been observed that the proposed FFT-OFDM system with FRF interleaving outperforms that with helical interleaving in terms of BER, as the FRF interleaver has better randomization capabilities than the helical interleaver. On the other hand, the proposed DWT-OFDM system with the FRF interleaver provides approximately the same BER performance as that of the DWT-OFDM system with the helical interleaver. As a result, our recommendation is to use the proposed FRF interleaver for the FFT-OFDM system. Finally, the proposed FRF interleaver is more complex than the helical interleaver, but the FRF interleaver can improve the throughput and energy efficiency of critical systems, where their power consumption is essential, compared with the others. The simulation results have shown that the performances of the proposed FRF and helical interleavers have noticeable improvements compared to that of the conventional (FFT/DWT)-OFDM. Finally, the proposed FRF interleaver is more complex than the helical interleaver. One possible future focus of this paper is a deep investigation of the performance of conventional interleavers compared to our proposed FRF interleaver.

Author Contributions: Conceptualization, F.H.E.-F. and R.A.R.; methodology, F.H.E.-F. and R.A.R.; software, F.H.E.-F.; validation, F.H.E.-F. and R.A.R.; formal analysis, F.H.E.-F. and F.E.A.E.-S.; investigation, F.H.E.-F. and M.K.; resources, A.J.A. and J.S.A.; writing—original draft preparation, F.H.E.-F. and R.A.R.; writing—review and editing, F.H.E.-F., R.A.R., A.J.A. and J.S.A.; visualization, F.H.E.-F. and F.E.A.E.-S.; supervision, F.H.E.-F. and R.A.R.; project administration, F.H.E.-F.; funding acquisition, R.A.R., A.J.A., M.K. and J.S.A. All authors have read and agreed to the published version of the manuscript.

Funding: This research was funded by the Scientific Research Deanship at the University of Ha'il, Saudi Arabia, through project number RG-20019.

Conflicts of Interest: The authors declare no conflict of interest.

References

1. Amasa, R. Inter Carrier Interference Cancellation in OFDM Systems. Master's Thesis, Department of Electronics and Communication Engineering, National Institute of Technology Rourkela, Rourkela, India, 2009.
2. Bingham, J. Multicarrier modulation for data transmission: An idea whose time has come. *IEEE Commun. Mag.* **1990**, *28*, 5–14. [CrossRef]
3. Kiani, A.; Baghersalimi, G.; Zanj, B. Performance Assessment of DFT-OFDM and DWT-OFDM Systems in the Presence of the HPA Nonlinearity. In Proceedings of the IEEE 11th International Conference on Telecommunications, ConTEL 2011, Graz, Austria, 15–17 June 2011; pp. 273–278.
4. Asif, R.; Abd-Alhameed, R.; Anoh, O.; Hussaini, A.S.; Rodriguez, J. Performance Comparison between DWT-OFDM and FFT-OFDM Using Time Domain Zero Forcing Equalization. In Proceedings of the 2012 International Conference on Telecommunications and Multimedia (TEMU 2012), Crete, Greece, 30 July–1 August 2012.
5. Bernardo, L.; Lopes, P. Performance of Chaotic Modulation Over Rayleigh Fading Channels Using FFT-OFDM and DWT-OFDM. In Proceedings of the 20th European Signal Processing Conference, Bucharest, Romania, 27–31 August 2012; pp. 729–733.
6. Kol, V.K.; Mishra, A. Discrete wavelet transform based OFDM-IDMA system with AWGN channel. In Proceedings of the Students Conference on Engineering and Systems (SCES), Allahabad, India, 12–14 April 2013.
7. Patel, V.V.; Patil, R.N. Minimization of PAPR in OFDM system using IDWT/DWT, clipping and filtering combined with huffman coding method. In Proceedings of the IEEE International Conference on Emerging Trends in Computing, Communication and Nanotechnology (ICE-CCN), Tirunelveli, India, 25–26 March 2013; pp. 250–254.
8. Meenakshi, D.; Prabha, S.; Raajan, N.R. Compare the performance analysis for FFT based MIMO-OFDM with DWT based MIMO-OFDM. In Proceedings of the IEEE International Conference on Communication and Signal Processing (ICCSP'13), Melmaruvathur, Tamilnadu, India, 3–5 April 2013; pp. 441–445.
9. Sandberg, S.D.; Tzannes, M.A. Overlapped discrete multitone modulation for high speed copper wire communications. *IEEE J. Sel. Areas Commun.* **1995**, *13*, 1571–1585. [CrossRef]
10. Negash, B.G.; Nikookar, H. Wavelet based OFDM for wireless channels. In Proceedings of the Vehicular Technology Conference, Atlantic City, NJ, USA, 7–11 October 2001.
11. Akansu, A.N.; Lin, X. A comparative performance evaluation of DMT (OFDM) and DWMT (DSBMT) based DSL communications systems for single and multitone interference. In Proceedings of the IEEE International Conference on Acoustics, Speech and Signal Processing, Seattle, WA, USA, 15 May 1998.
12. Ahmed, N. Joint Detection Strategies for Orthogonal Frequency Division Multiplexing. Master's Thesis, Electrical and Computer Engineering Department, Rice University, Houston, TX, USA, 2000; pp. 1–51.
13. Sung, C.K.; Chung, S.-Y.; Heo, J.; Lee, I. Adaptive Bit-Interleaved Coded OFDM with Reduced Feedback Information. *IEEE Trans. Commun.* **2007**, *55*, 1649–1655. [CrossRef]
14. Lei, S.W.; Lau, V.K.N. Performance Analysis of Adaptive Interleaving for OFDM Systems. *IEEE Trans. Veh. Technol.* **2002**, *51*, 435–444.
15. Wang, Y.; Zhu, Q.F. Error control and concealment for video communication: A review. *Proc. IEEE* **1998**, *86*, 974–997. [CrossRef]
16. Padmaja, C.; Malleswari, B.L. Performance analysis of Adaptive Bit-interleaved Coded Modulation in OFDM using Zero Padding Scheme. *IOSR J. Mob. Comput. Appl. (IOSR-JMCA)* **2015**, *2*, 2394-0050.
17. Chaturvedi, V.; Gupta, V.K.; Dehradun, D.I.T. Performance Analysis for Different Interleavers in Various Modulation Schemes with OFDM over an AWGN Channel. *IOSR J. Eng.* **2016**, *2*, 760–767. [CrossRef]
18. Harada, H.; Prasad, R. *Simulation and Software Radio for Mobile Communications*; Artech House: Norwood, MA, USA, 2002.
19. Andrews, J.G.; Ghosh, A.; Muhamed, R. *Fundamentals of WiMAX*; Pearson Education, Inc.: New York, NY, USA, 2007.
20. Webb, W. *Wireless Communications: The Future*; John Wiley & Sons Ltd.: Hoboken, NJ, USA, 2007.
21. Sari, H.; Karam, G.; Jeanclaude, I. Transmission Techniques for Digital Terrestrial TV Broadcasting. *IEEE Commun. Mag.* **1995**, *33*, 100–109. [CrossRef]
22. Tsai, Y.; Zhang, G.; Pan, J.-L. Orthogonal Frequency Division Multiplexing with Phase Modulation and Constant Envelope Design. In Proceedings of the 2005 IEEE Military Communications Conference (MILCOM), Atlantic City, NJ, USA, 17–20 October 2005; Volume 4, pp. 2658–2664.

23. Hara, S.; Prasad, R. Overview of multicarrier CDMA. *IEEE Commun. Mag.* **1997**, *35*, 126–144. [CrossRef]
24. Hassa, E.S.; Zhu, X.; El-Khamy, S.E.; Dessouky, M.I.; El-Dolil, S.A.; Abd El-Samie, F.E. Enhanced Performance of OFDM and Single-Carrier Systems Using Frequency Domain Equalization and Phase Modulation. In Proceedings of the 26th National Radio Science Conference (NRSC2009), Cairo, Egypt, 17–19 March 2009; pp. 1–10.
25. Strang, G.; Nguyen, T. *Wavelets and Filter Banks*; Wellesly-Cambridge Press: Wellesley, MA, USA, 1997.
26. Rappaport, T.S. *Wireless Communications: Principles & Practice*; Prentice Hall: Hoboken, NJ, USA, 1998.
27. Proakis, J.G. *Digital Communication*, 4th ed.; McGraw-Hill Education: New York, NY, USA, 2001.
28. Bansal, R.; Anand, S.; Purwar, D.; Shukla, A. Application of Antenna Diversity in HIDMA scheme using gold codes. In Proceedings of the IEEE 2011 International Conference on Communication Systems and Network Technologies, Katra, India, 3–5 June 2011; pp. 228–231.
29. Hao, D.; Hoeher, P.A. Helical Interleaver Set Design for Interleave-Division Multiplexing and Related Techniques. *IEEE Commun. Lett.* **2008**, *12*, 843–845. [CrossRef]
30. Ramabadran, S.; Madhukumar, A.S.; Wee Teck, N.; See, C.M.S. Parameter Estimation of Convolutional and Helical Interleavers in a Noisy Environment. *IEEE Access* **2017**, *5*, 6151–6167. [CrossRef]
31. Iqbal, S.; Gupta, P.; Kumar, S. Performance Analysis of Digital Video Broadcasting—Cable (DVB-C) System with 64-QAM Modulation and Interleaving Schemes. *Int. J. Innov. Res. Electr. Electron. Instrum. Control Eng.* **2021**, *9*, 65–68.
32. Asghar, R. Flexible Interleaving Subsystems for FEC in Baseband Processors. Ph.D. Thesis, Linköping University, Linköping, Sweden, 2010; p. 189.
33. Juliet, A.M.; Jayashri, S. Concert Investigation of Novel Deterministic Interleaver for OFDM-IDMA system. *Indian J. Comput. Sci. Eng.* **2013**, *4*, 353–363.
34. Daubechies, I. *Ten Lectures on Wavelets*; Society for Industrial and Applied Mathematics: Philadelphia, PA, USA, 1992.
35. Jain, R. Channel Models a Tutorial. 2007. Available online: http://www.cse.wustl.edu/~jain/cse574-08/ftp/channel_model_tutorial.pdf (accessed on 1 December 2021).

applied sciences

MDPI

Article

Novel MIMO Antenna System for Ultra Wideband Applications

Abdullah G. Alharbi [1], Umair Rafique [2,*], Shakir Ullah [3], Salahuddin Khan [4], Syed Muzahir Abbas [5], Esraa Mousa Ali [6], Mohammad Alibakhshikenari [7] and Mariana Dalarsson [8,*]

1 Department of Electrical Engineering, Faculty of Engineering, Jouf University, Sakaka 42421, Saudi Arabia; a.g.alharbi@ieee.org
2 Department of Information Engineering, Electronics, and Telecommunications, Sapienza University of Rome, 00184 Rome, Italy
3 Telecommunication Engineering Department, University of Engineering & Technology, Mardan 23200, Pakistan; shakirhayat.eng@gmail.com
4 College of Engineering, King Saud University, P.O. Box 800, Riyadh 11421, Saudi Arabia; drskhan@ksu.edu.sa
5 Faculty of Science and Engineering, School of Engineering, Macquarie University, Sydney, NSW 2109, Australia; syed.abbas@mq.edu.au
6 Faculty of Aviation Sciences, Amman Arab University, Amman 11953, Jordan; esraa_ali@aau.edu.jo
7 Department of Signal Theory and Communications, Universidad Carlos III de Madrid, 28911 Leganés, Madrid, Spain; mohammad.alibakhshikenari@uc3m.es
8 School of Electrical Engineering and Computer Science, KTH Royal Institute of Technology, SE 100-44 Stockholm, Sweden
* Correspondence: umair.rafique@uniroma1.it (U.R.); mardal@kth.se (M.D.)

Abstract: The design of a 4 × 4 MIMO antenna for UWB communication systems is presented in this study. The single antenna element is comprised of a fractal circular ring structure backed by a modified partial ground plane having dimensions of 30 × 30 mm^2. The single antenna element has a wide impedance bandwidth of 9.33 GHz and operates from 2.67 GHz to 12 GHz. Furthermore, the gain of a single antenna element increases as the frequency increases, with a peak realized gain and antenna efficiency of 5 dBi and >75%, respectively. For MIMO applications, a 4 × 4 array is designed and analyzed. The antenna elements are positioned in a plus-shaped configuration to provide pattern as well as polarization diversity. It is worth mentioning that good isolation characteristics are achieved without the utilization of any isolation enhancement network. The proposed MIMO antenna was fabricated and tested, and the results show that it provides UWB response from 2.77 GHz to over 12 GHz. The isolation between the antenna elements is more than 15 dB. Based on performance attributes, it can be said that the proposed design is suitable for UWB MIMO applications.

Keywords: 4 × 4 MIMO; UWB communication; fractal circular ring; pattern and polarization diversity; plus shape configuration

Citation: Alharbi, A.G.; Rafique, U.; Ullah, S.; Khan, S.; Abbas, S.M.; Ali, E.M.; Alibakhshikenari, M.; Dalarsson, M. Novel MIMO Antenna System for Ultra Wideband Applications. *Appl. Sci.* **2022**, *12*, 3684. https://doi.org/10.3390/app12073684

Academic Editor: Amalia Miliou

Received: 20 February 2022
Accepted: 4 April 2022
Published: 6 April 2022

1. Introduction

Recent progress in wireless communication necessitates a system capable of sharing massive amounts of data quickly with increased capacity, dependability, and security with less complexity. Ultra wideband (UWB) technology can meet these expectations since it allows high data rates while employing low-cost infrastructure [1,2]. The traditional UWB technology, on the other hand, has problems with multipath propagation. To mitigate this effect, multiple-input multiple-output (MIMO) technology has been introduced. Furthermore, one of the most promising solutions for meeting the needs of UWB systems is to design MIMO antennas. It is a difficult task to design such an antenna because it adds extra performance characteristics to consider. Antenna element isolation is one of the most essential performance characteristics, which should be as low as possible to avoid channel capacity loss.

Several configurations for improving isolation have been published in the literature. Designing stubs with the ground plane or the radiating element helps improve isolation between MIMO antenna elements. In [3], a spatial diversity-based MIMO antenna was designed for UWB applications. To obtain a wide impedance bandwidth, a triangular-shaped patch radiator with a modified ground plane was used. The radiators do not radiate directly towards each other, which can lead to better isolation. In the ground plane, inverted L-shaped stubs and a complementary split ring resonance (CSRR) were designed to improve isolation even further. The authors in [4] utilized the same kind of configuration. In this design, a tapered fed polygon-shaped patch radiator was designed to achieve UWB response. The authors in [5] used the same type of isolation enhancement technique as presented in [3,4]. They designed a co-planar waveguide (CPW)-fed beveled-shaped patch radiator to achieve a UWB response. In addition, the radiating structure was composed of two half-wavelength rectangular single CSRRs (RSCSRRs) of different dimensions to achieve band notch characteristics for Worldwide Interoperability for Microwave Access (WiMAX) and Wireless Local Area Network (WLAN) frequency bands. In [6], an eye-shaped slot radiator-based MIMO antenna design was presented. The ground plane was developed with an extruded T-shaped stub to improve isolation. This simple configuration tends to achieve an isolation of >20 dB and ECC < 0.02. In [7], a 2-element dual-ring-based MIMO antenna was designed. The main patch radiator consists of a ring element, while the second ring behaves as a parasitic element. To improve the isolation between the antennas, two inverted U-stubs were designed into the ground plane.

The employment of electromagnetic bandgap (EBG) structures, metamaterials, and frequency-selective surfaces (FSSs) to increase MIMO antenna element isolation has been proposed in the literature as a viable technology. In [8], a modified radiator with a FSS was reported. The design was non-planar, with four elements placed around a cube of polystyrene. A quad-port cognitive radio MIMO antenna was presented in [9]. The single antenna consists of a circular monopole geometry and a $\lambda/4$ monopole for the notch band. For high isolation, multiple EBG cells having different dimensions and an isolating stub between the UWB and notch band radiators were employed. A two-port MIMO antenna with improved E-plane isolation was presented in [10]. A meandered EBG structure was designed on both sides of the substrate to improve isolation. In [11], a MIMO antenna design was presented having arrow shaped radiator. The presented design operates within the UWB spectrum with three notch bands obtained by cutting the slots within the radiator. To improve the isolation performance, a T-shaped stub was used with the ground plane, and two EBG cells were placed on each side of the feeding line. Additionally, the size of the structure is quite large for a two-port configuration. In [12], a miniaturized UWB MIMO antenna design was presented. The MIMO antenna elements were composed of a combination of rectangle and semi-circular arc fed using a stepped microstrip fee line. For improved isolation between MIMO elements, a Jerusalem cross-shaped FSS structure and a meandered-shaped decoupling structure were designed. All the above-reported designs have EBG/FSS structures, which are complex to fabricate as they have a "via" that connects the top and ground plane and are large in size.

The use of neutralization lines or decoupling structures between the radiators can also provide high isolation. In [13], a decoupling strip was diagonally placed between the antenna elements for isolation enhancement. In addition, a dumbbell-shaped stub was designed with the ground plane, and the same was placed on the top side of the substrate. From the presented configuration, an isolation of >15 dB was achieved in the frequency range of 3–11 GHz. In [14], a four-port dual-band nocth UWB MIMO antenna was presented. Four rhombic-shaped patch radiators were arranged in a polarization diversity configuration. Two CSRRs were etched from the radiators to reject WLAN and WiMAX frequency bands. A plus-shaped metal strip was designed with the ground plane for enhanced isolation. In [15], a design for a compact super wideband (SWB) antenna was presented. The presented SWB radiator geometry consists of a combination of circular, rectangular, and trapezoidal shapes. The bandwidth of the antenna was enhanced by placing slots in the radiator and by a modified ground plane. Different MIMO configurations were analyzed, such as spatial, pattern, and polarization diversity with a common ground plane. From the presented configurations, an isolation of ≥30 dB was achieved. In [16], an octagonal band notch UWB MIMO antenna design was demonstrated. Two hexagonal-shaped CSRRs (HCSRRs) were etched from the octagonal radiator to achieve band stop characteristics around the WLAN and WiMAX frequency bands. A decoupling structure consisting of three thin metal strips was designed in the middle of the radiation elements to achieve an isolation of >18 dB in the UWB frequency range. The same kind of isolation enhancement technique was described in [17]. In [18], a simple polarization diversity-based four-element MIMO configuration was presented for UWB applications. The single antenna element consists of a combination of rectangular and circular patch radiators. In [19], an eight-element semi-circular arc-based MIMO antenna design was presented for UWB applications. The UWB response was achieved by etching a semi-elliptical slot in the ground plane and designing an inverted L-shaped strip in conjunction with the ground plane. Initially, the authors realized a four-element MIMO antenna design placed in a plus-shaped configuration. After that, an eight-element MIMO antenna configuration was demonstrated for UWB applications. Thin metallic strips were designed between the radiators to achieve an isolation of >15 dB in the frequency range of 2.84–11 GHz.

In the literature, some researchers utilized defected ground structures (DGSs) to improve the isolation between MIMO antennas. A two-element aperture-coupled complementary Sierpinski gasket equilateral triangular fractal MIMO antenna was designed for UWB performance in [20]. For enhanced isolation, a complementary Archimedean spiral-based DGS was utilized. The use of DGS led to an isolation level of ≥15.8 dB. In [21], for UWB applications, a MIMO antenna with good isolation was demonstrated. For MIMO antenna design, slot elements having quasi F-shaped radiators and L-shaped open-slots were adopted. In the ground plane, a decoupling structure consisting of a thin slot and a fork-shaped slot was implemented to increase isolation between the radiators and in the lower band.

The techniques presented above provide better isolation between antenna elements, but they are quite complex and their fabrication is difficult. In this paper, a novel MIMO antenna configuration is presented for UWB applications. The antenna elements are positioned in a plus-shaped arrangement to provide pattern and polarization diversity [19,22]. It is observed from the presented results that the designed MIMO antenna operates from 2.77–12 GHz. Furthermore, the presented MIMO antenna did not utilize any isolation enhancement technique, which makes it simple and it can easily be fabricated through low-cost fabrication techniques.

2. Proposed MIMO Configuration

2.1. Mimo Single Element

The design of a single antenna element is depicted in Figure 1. The front side of the designed antenna consists of a fractal circular ring radiator [23] fed using a 50 Ω microstrip feed line (see Figure 1a), while the back side consists of a notch loaded trapezoidal partial ground plane, as illustrated in Figure 1b. The radius R_1 of the outer circular ring is calculated using the design expressions mentioned in [24], and the width of the feeding line is calculated using the procedure described in [25]. From Figure 1a, one can observe that six circular rings of radius R_3 are embedded in the main radiator to realize a fractal geometry. These rings are rotated at an angle of 60°. Furthermore, the trapezoidal ground plane is utilized (see Figure 1b) to create wideband multi-resonance characteristics of the input impedance, which leads to achieving enhanced bandwidth [26], while the square notch etched from the ground plane offers maximum impedance matching in the operating bandwidth. The proposed antenna is designed using a low-cost FR-4 laminate with ϵ_r = 4.3 and h = 1.6 mm. The overall design dimensions of the antenna are: W_S = 30, L_S = 30, R_1 = 9.5, R_2 = 7.5, R_3 = 3.5, R_4 = 2.4, W_f = 3, L_f = 10, W_g = 30, L_g = 9.5, L_{g1} = 5.5, and S = 3 (all dimensions are in mm).

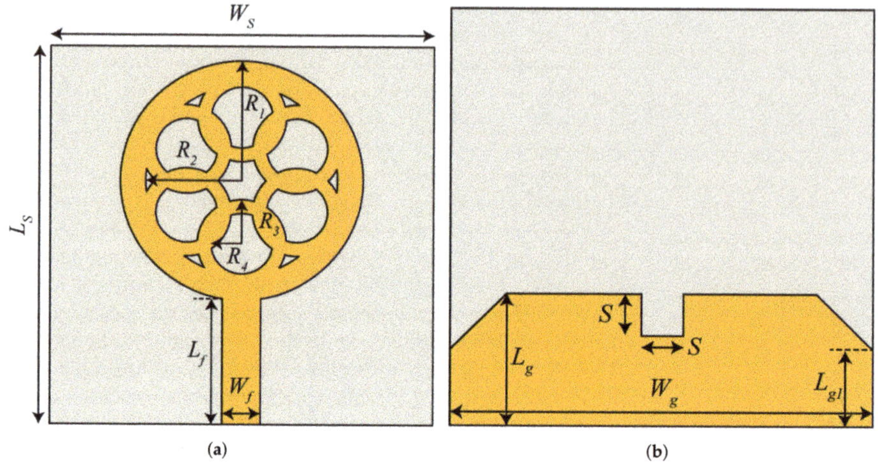

Figure 1. Schematic of the proposed single antenna element (**a**) front side (**b**) back side.

The construction of the proposed antenna starts with the design of a circular ring patch radiator backed by a conventional partial ground plane, shown in Figure 2a, and its respective reflection coefficient (S_{11}) is depicted in Figure 3. From the results, it is observed that the circular ring provides a dual-band response around 3 GHz and 10 GHz. To excite more resonant modes in the band of interest, six circular rings of radius R_3 are embedded in the main radiator (see Figure 2b), which ultimately leads to a fractal geometry. This configuration increased the impedance bandwidth, but there is still a mismatch between 7.58 GHz and 9.38 GHz (see Figure 3). To improve the impedance matching, a trapezoidal shaped ground plane with a square notch is utilized, as shown in Figure 2c. The modification in the ground structure leads to achieving an UWB frequency response ranging from 2.69 GHz to 12 GHz, as shown in Figure 3.

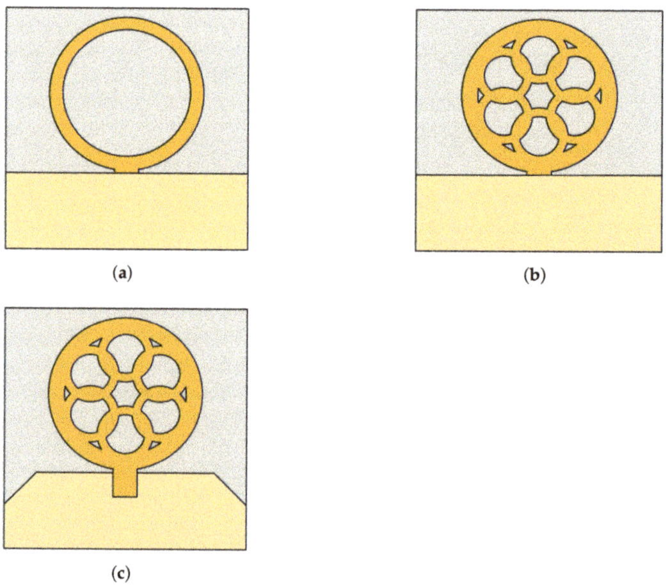

Figure 2. Design evolution of single antenna element. (**a**) Step-1. (**b**) Step-2. (**c**) Step-3 (Proposed).

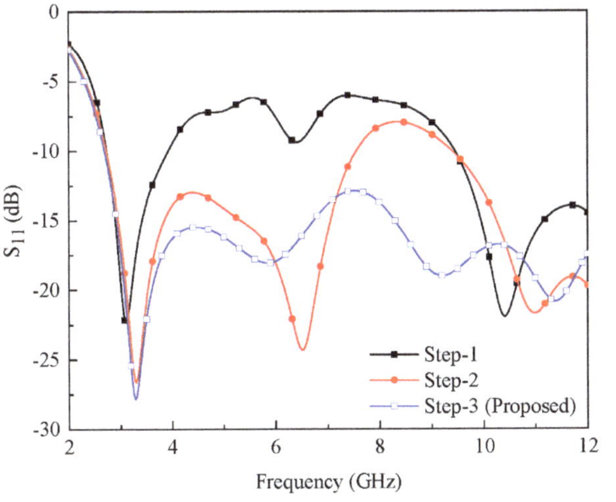

Figure 3. Reflection coefficients (S_{11}) of different design stages.

Figure 4 depicts the presented antenna element's realized gain and radiation efficiency. The antenna gain varies in the range of 0.55–5 dBi in the operating bandwidth, while the radiation efficiency is noted to be >75% in the band of interest (see Figure 4).

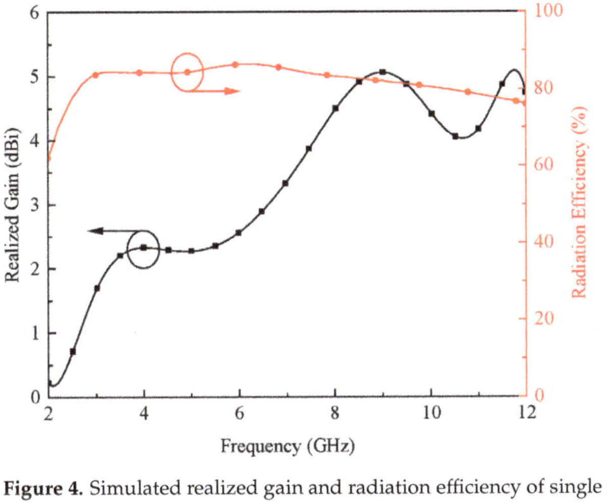

Figure 4. Simulated realized gain and radiation efficiency of single antenna element.

2.2. 4 × 4 MIMO Design

The proposed MIMO antenna configuration is shown in Figure 5. From the figure, one can observe that four antenna elements are arranged in a plus-shaped configuration, which tends to achieve both pattern and polarization diversity. Antenna-1 and antenna-2 are placed at a 90° angle, which is the typical configuration of polarization diversity, while antenna-1 and antenna-3 offer pattern diversity (see Figure 5). As a whole, each antenna element is able to provide both pattern and polarization diversity characteristics. It is also worth mentioning that no decoupling network is used to enhance the isolation between antenna elements. The dimensions of the antenna elements are the same as those of a single element (see Figure 1). As shown in Figure 5a, there is an extra space created between the antenna elements, which is represented by $W_{S1} \times L_{S1}$ = 30 × 30 mm^2. Therefore, the overall dimensions of the proposed MIMO antenna are 90 × 90 mm^2 (consider both sides of the plus shape).

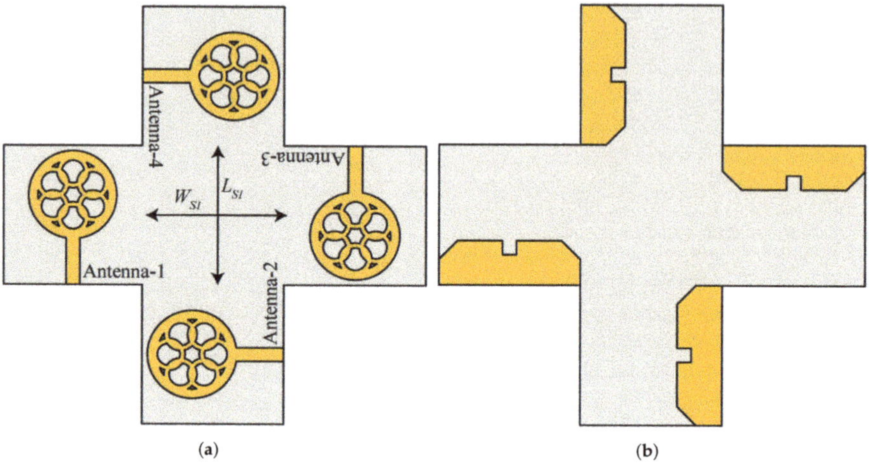

Figure 5. Configuration of the proposed UWB MIMO antenna (**a**) front side (**b**) back side.

Computer Simulation Technology (CST) Microwave Studio 2021 was used to develop and simulate the proposed MIMO antenna. The simulated S-parameters of the designed MIMO are shown in Figure 6. In the figure, the reflection coefficients are represented by S_{ii} ($i \in 1, 2, 3, 4$), while the isolation between adjacent elements is denoted by S_{ji} ($i, j \in 1, 2, 3, 4$ and $i \neq j$), and for parallel elements, it is represented by S_{ki} ($k, i \in 1, 2, 3, 4$ and $|k - i| = 2$). The MIMO antenna's impedance bandwidth is 9.27 GHz in the range of 2.73–12 GHz. Furthermore, the isolation between adjacent antenna elements, denoted by S_{ji}, is >17 dB whereas the isolation between parallel elements, denoted by S_{ki}, is ≥14 dB.

To better understand the isolation characteristics, the surface current distribution (see Figure 7) is plotted by exciting only port-1 of the array, while the rest of the ports are terminated with a 50 Ω matched load. Three different frequencies, such as 3 GHz, 6 GHz, and 9 GHz, are used to depict the surface current. It can be observed from Figure 7 that the surface current generated by antenna-1 does not influence adjacent or parallel placed elements.

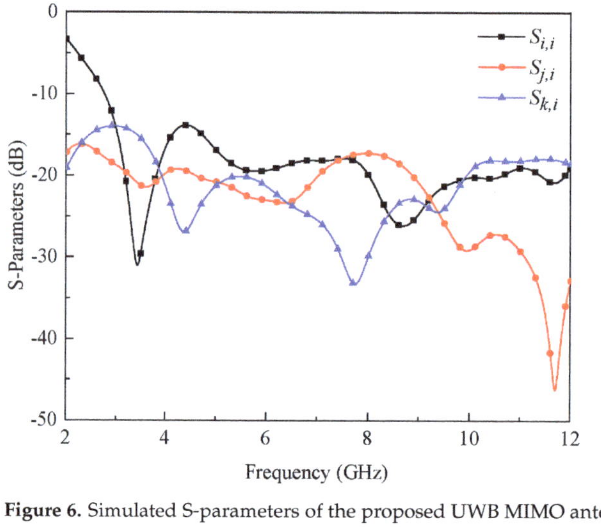

Figure 6. Simulated S-parameters of the proposed UWB MIMO antenna.

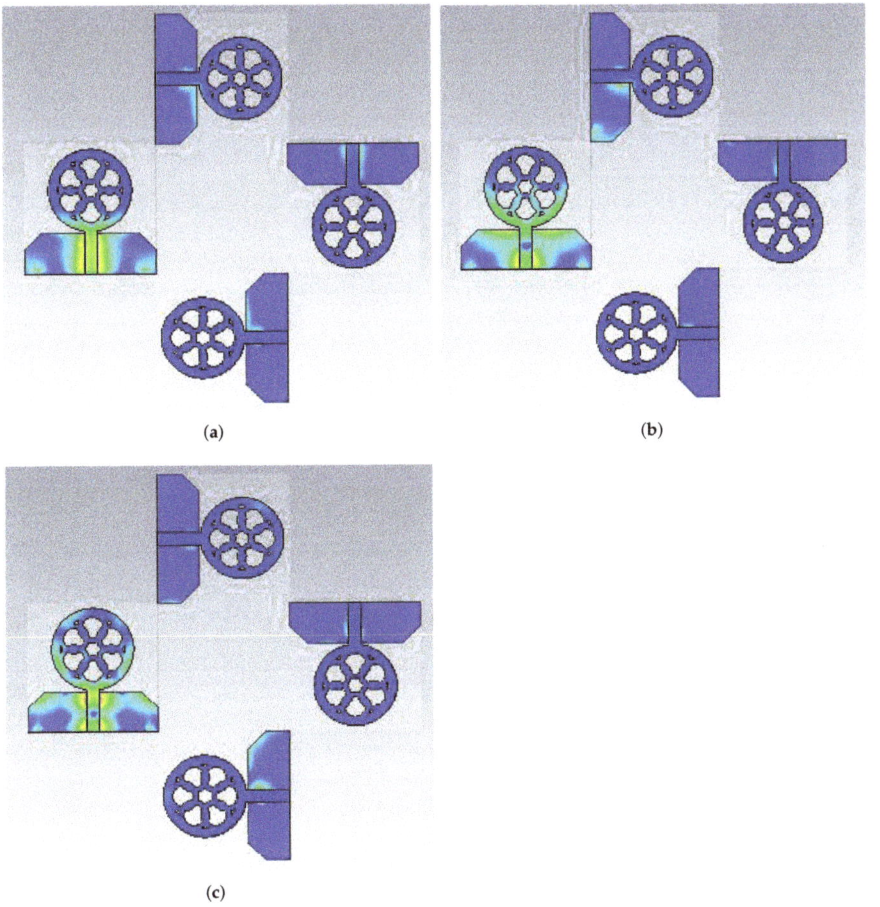

Figure 7. Surface current distribution of the proposed UWB MIMO antenna when port-1 is excited. (**a**) 3 GHz. (**b**) 6 GHz. (**c**) 9 GHz.

3. Fabrication and Measured Results

A prototype of the four-element MIMO antenna was fabricated (see Figure 8) to verify the simulations. For measurement purposes, a Precision Network Analyzer (PNA) E8363C by Agilent Technologies is utilized. The measured S-parameters are shown in Figure 9. Due to symmetry, the reflection coefficients of antenna-1 and antenna-2 are presented. From Figure 9, it is evident that the measured S-parameters matched well with the simulated ones (see Figure 6). The measured impedance bandwidth is 9.23 GHz in the frequency range of 2.77–12 GHz. As demonstrated in Figure 9b, the observed isolation is >15 dB for both neighbouring and parallel antenna elements. The discrepancies between the results are due to SMA connector losses and fabrication intolerances.

Figures 10–13 demonstrate the radiation parameters of the proposed MIMO antenna for all ports at frequencies of 3 GHz, 6 GHz, and 9 GHz. In case of port-1, for 3 GHz (see Figure 10a), omnidirectional pattern is observed for xz-plane ($\phi = 0°$), while for yz-plane ($\phi = 90°$), antenna exhibits typical monopole like pattern. For 6 GHz, a quasi-omnidirectional pattern is noted for xz-plane and a monopole like pattern is observed for yz-plane, as shown in Figure 10b. For higher frequencies, such as 9 GHz, a quasi-omnidirectional pattern with some ripples is observed for both the planes, as illustrated in Figure 10c. For port-2, the same kinds of patterns are observed, but in this case, antenna-1

xz-plane radiation pattern is equal to antenna-2's yz-plane radiation pattern, and vice ver This effect can clearly be observed from the patterns in Figure 11. From Figure 12, one c observe that antenna-3 offers the same radiation characteristics as antenna-1. In this cas the yz-plane of antenna-3 provides pattern diversity for the reported frequency bands. Th same is the case with antenna-4, whose yz-plane offers pattern diversity in comparison t antenna-2's characteristics. As a whole, from the results of Figures 10–13, one can observe that all the antenna elements have the ability to exhibit pattern and polarization diversity. Furthermore, from Figures 10–13, one can also observe that the simulated and measured radiation characteristics are in good agreement. Some discrepancies are observed between the simulated and measured data, especially at higher frequencies (9 GHz), which could possibly arise due to fabrication tolerances at higher frequencies and far-field measurement setup losses.

(**a**) (**b**)

Figure 8. Fabricated prototype of the proposed UWB MIMO antenna (**a**) front side (**b**) back side.

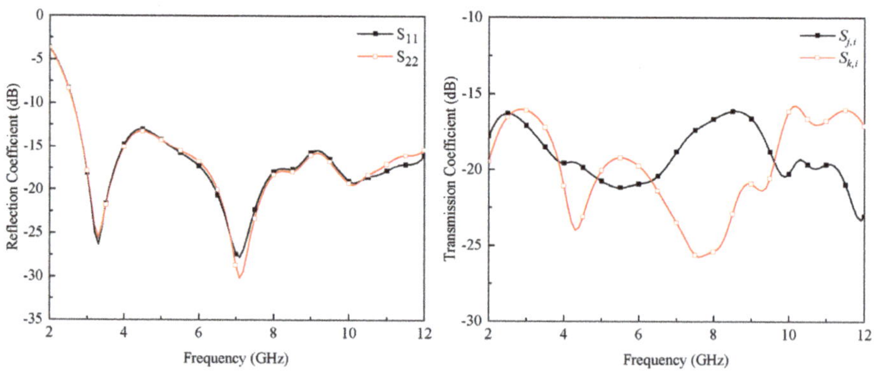

(**a**) (**b**)

Figure 9. Measured (**a**) reflection and (**b**) transmission coefficients of the proposed UWB MIMO antenna.

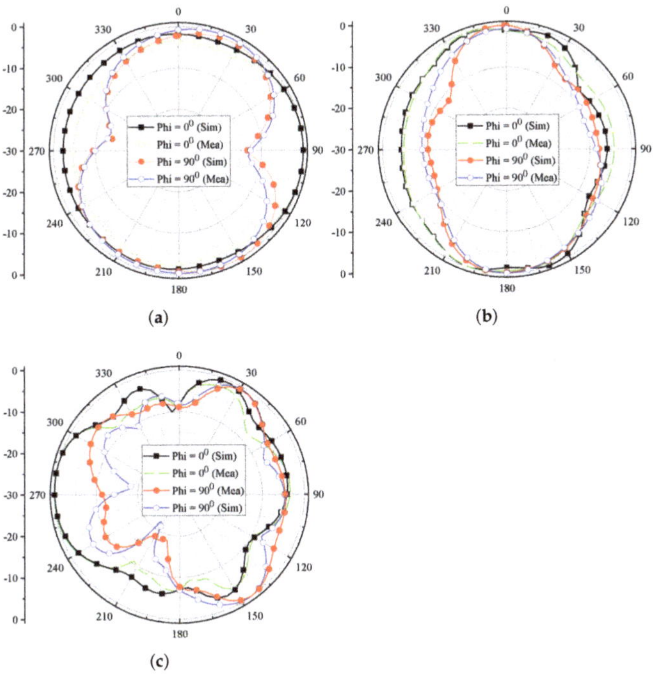

Figure 10. Radiation characteristics of the proposed UWB MIMO for port-1. (**a**) 3 GHz. (**b**) 6 GHz. (**c**) 9 GHz.

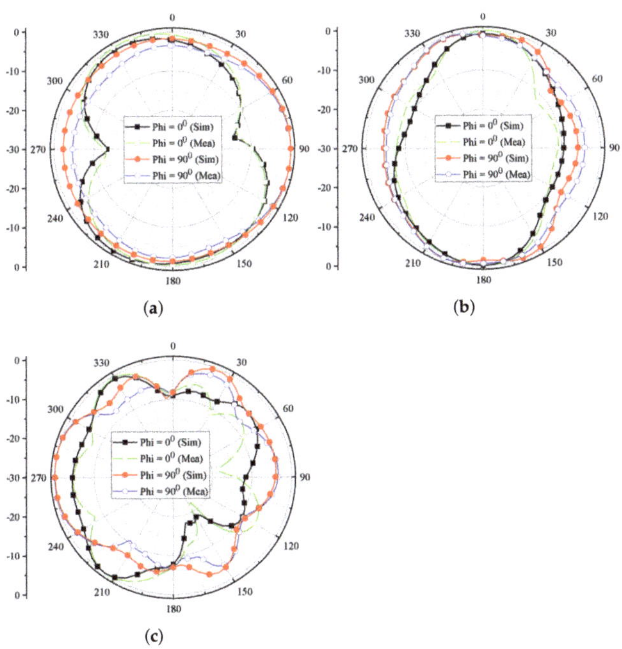

Figure 11. Radiation characteristics of the proposed UWB MIMO for port-2. (**a**) 3 GHz, (**b**) 6 GHz. (**c**) 9 GHz.

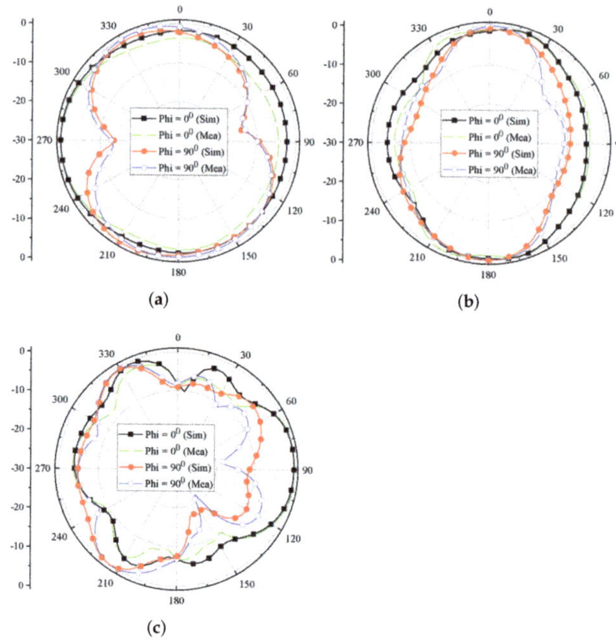

Figure 12. Radiation characteristics of the proposed UWB MIMO for port-3. (**a**) 3 GHz. (**b**) 6 GHz. (**c**) 9 GHz.

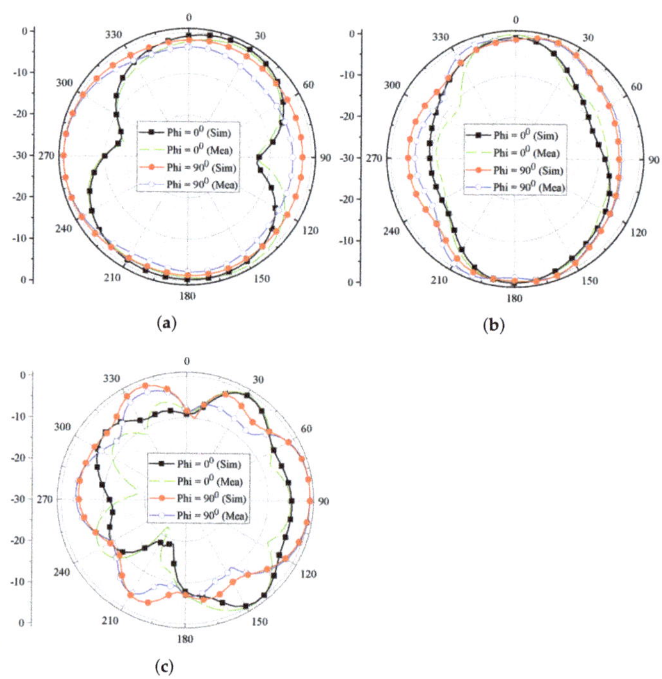

Figure 13. Radiation characteristics of the proposed UWB MIMO for port-4. (**a**) 3 GHz. (**b**) 6 GHz. (**c**) 9 GHz.

4. Diversity Performance Parameters

4.1. Envelope Correlation Coefficient (ECC)

To evaluate the proposed MIMO antenna's diversity performance, ECC is analyzed and presented in Figure 14. The far-field characteristics are used to determine the ECC as [27,28]:

$$ECC = \left| \frac{\iint_{4\pi} S_i(\theta,\phi).S_j^*(\theta,\phi)d\Omega}{\sqrt{\iint_{4\pi} S_i(\theta,\phi).S_i^*(\theta,\phi)d\Omega \iint_{4\pi} S_j(\theta,\phi).S_j^*(\theta,\phi)d\Omega}} \right|^2 \quad (1)$$

where S_i and S_j represents far-field radiation characteristics of port i and port j.

According to [29], the value of ECC should be set at <0.5 for practical applications. As shown in Figure 14, the proposed MIMO antenna exhibits ECC < 0.1.

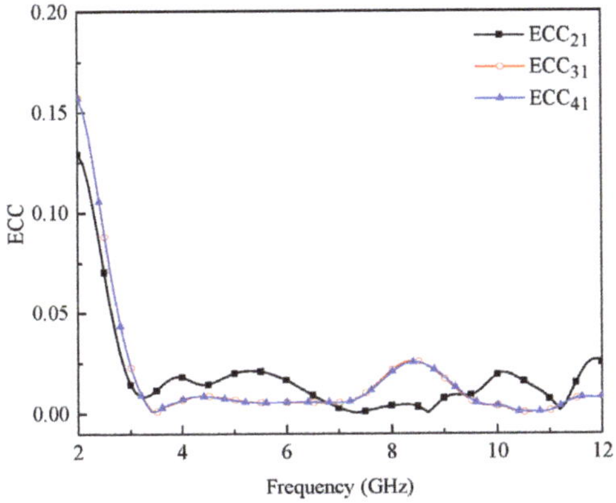

Figure 14. ECC of the proposed UWB MIMO antenna.

4.2. Diversity Gain (DG)

Another important feature that needs to be assessed is DG. The DG of the MIMO antennas should be high (≈10 dB) in the operating bandwidth. It is expressed in terms of ECC as [28]:

$$DG = 10\sqrt{1 - ECC^2} \quad (2)$$

Figure 15 depicts the DG of the proposed MIMO antenna, and it is observed that it fluctuates around a 10 dB scale for the entire operating bandwidth.

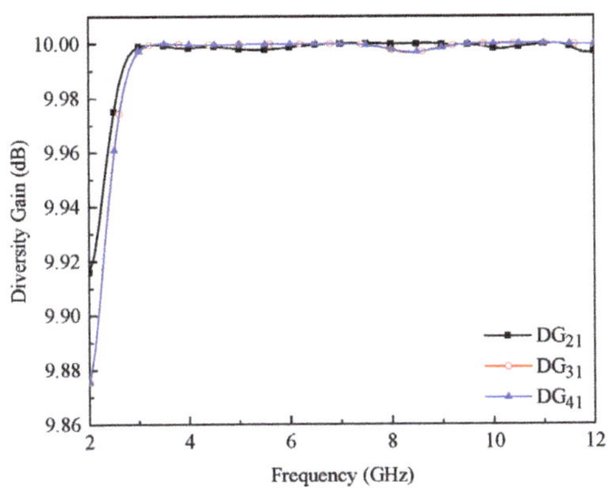

Figure 15. Diversity gain of the proposed UWB MIMO antenna.

4.3. Mean Effective Gain (MEG)

In comparison to an isotropic antenna, MEG is a measurement of how much powe the antenna elements receive. The S-parameters of the MIMO antenna system can be usec to calculate the MEG as [22,28]:

$$MEG_i = 0.5\left[1 - |S_{nn}|^2 - |S_{nm}|^2\right] \quad (3)$$

and

$$MEG_j = 0.5\left[1 - |S_{nm}|^2 - |S_{mm}|^2\right] \quad (4)$$

For the proposed MIMO antenna, the MEG is predicted between port-1 & port-2 and port-1 & port-3 (see Figure 16). It can be noted that the MEG of the ports is ≤ -3 dB, while the ratios of MEG1/MEG2 and MEG1/MEG3 are less than 3 dB.

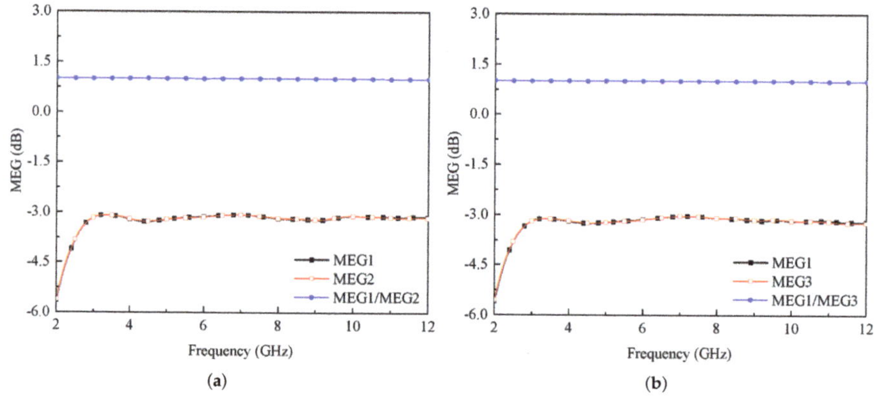

Figure 16. MEG between (**a**) port-1 & port-2 and (**b**) port-1 & port-3.

A comparison among proposed and previously presented four-port UWB MIMO antenna designs is presented in Table 1. The proposed design offers high impedance bandwidth compared to the designs presented in [8,12,13,19]. Although the dimensions of the proposed MIMO antenna are large, it provides comparable isolation and ECC characteristics without the utilization of any isolation enhancement network compared to the designs listed in Table 1. Moreover, the design of [18] did not utilize any isolation enhancement network, so it offers a high value of ECC compared to the proposed design. In addition, the authors in [8,13,14,16,19] calculated ECC using S-parameters, which is not recommended for practical applications.

Table 1. Comparison among proposed and previously published four-port UWB MIMO antennas.

Ref.	Array Size (mm^2)	Isolation Enhancement Network	Frequency Band (GHz)	Bandwidth (GHz)	Isolation (dB)	ECC	DG (dB)
[8]	32 × 36	FSS	3–10	7	>16	<0.0025	–
[12]	45 × 45	FSS + Decoupling Structure	3.1-10.6	7.5	>10	<0.17	>9.92
[13]	43 × 40	Decoupling Structure	3–11	8	>15	<0.2	>9.75
[14]	72 × 72	Neutralization Lines	2.8–13.3	10.5	>18	< 0.06	–
[16]	58 × 58	Decoupling Structure	3–16	13	>18	<0.07	–
[18]	40 × 40	Not utilized	3–13.5	10.5	>15	<0.4	>9.9
[19]	92 × 92	Neutralization Lines	2.84–11	8.16	>15	<0.02	–
This work	90 × 90	Not utilized	2.77–12	9.23	>15	<0.1	>9.97

5. Conclusions

A four-port MIMO antenna is designed for UWB applications. The array's single element consists of a fractal circular ring patch radiator backed by a square notch loaded trapezidal-shaped partial ground plane. The results show that the designed single antenna element resonates from 2.67 GHz to over 12 GHz, and it offers a peak realized gain of 5 dBi and an antenna efficiency of >75%. Furthermore, a novel MIMO antenna configuration is designed to assess single element performance for UWB MIMO applications. To achieve pattern and polarization diversity, the MIMO antenna elements are placed in a plus-shaped configuration. According to the measured data, the designed MIMO antenna operates in the frequency range of 2.77–12 GHz and has a 9.23 GHz impedance bandwidth. In addition, isolation of >15 dB is noted between antenna elements with an ECC of <0.1, DG of ≈10 dB, and MEG of <3 dB.

Author Contributions: Conceptualization, A.G.A., U.R. and S.M.A.; methodology, U.R., S.M.A., E.M.A. and M.A.; software, U.R.; validation, A.G.A., U.R., S.U., S.K. and S.M.A.; formal analysis, A.G.A., U.R., S.M.A., M.A. and M.D.; investigation, A.G.A., S.M.A., E.M.A., M.D.; resources, S.K., E.M.A., M.A. and M.D.; writing—original draft preparation, U.R.; writing–review and editing, S.U., S.K., S.M.A., M.A. and M.D.; visualization, U.R.; supervision, S.M.A. and M.A.; project administration, A.G.A., S.K. and E.M.A.; funding acquisition, S.U., E.M.A., M.A. and M.D. All authors have read and agreed to the published version of the manuscript.

Funding: This work was funded by Universidad Carlos III de Madrid and the European Union's Horizon 2020 research and innovation programme under the Marie Sklodowska-Curie Grant 801538.

Acknowledgments: The authors would like to appreciate Universidad Carlos III de Madrid and the European Union's Horizon 2020 research and innovation programme for the funding of this research work under the Marie Sklodowska-Curie Grant 801538.

Institutional Review Board Statement: Not applicable.

Informed Consent Statement: Not applicable.

Data Availability Statement: Not applicable.

Conflicts of Interest: The authors declare no conflict of interest.

References

1. Lim, E.G.; Wang, Z.; Lei, C.U.; Wang, Y.; Man, K. Ultra wideband antennas: Past and present. *IAENG Int. J. Comput. Sci.* **2010**, *37*, 1–11.
2. Abbas, S.; Desai, S.; Esselle, K.; Volakis, J.; Hashmi, R. Design and characterization of a flexible wideband antenna using polydimethylsiloxane composite substrate. *Int. J. Antennas Propag.* **2018**, *2018*, 4095765. [CrossRef]
3. Khan, M.S.; Capobianco, A.D.; Asif, S.M.; Anagnostou, D.E.; Shubair, R.M.; Braaten, B.D. A compact CSRR-enabled UWB diversity antenna. *IEEE Antennas Wirel. Propag. Lett.* **2016**, *16*, 808–812. [CrossRef]
4. Chandel, R.; Gautam, A.K.; Rambabu, K. Tapered fed compact UWB MIMO-diversity antenna with dual band-notched characteristics. *IEEE Trans. Antennas Propag.* **2018**, *66*, 1677–1684. [CrossRef]
5. Bahmanzadeh, F.; Mohajeri, F. Simulation and fabrication of a high-isolation very compact MIMO antenna for ultra-wide band applications with dual band-notched characteristics. *AEU-Int. J. Electron. Commun.* **2021**, *128*, 153505. [CrossRef]
6. Chandel, R.; Gautam, A.K.; Rambabu, K. Design and packaging of an eye-shaped multiple-input–multiple-output antenna with high isolation for wireless UWB applications. *IEEE Trans. Components Packag. Manuf. Technol.* **2018**, *8*, 635–642. [CrossRef]
7. Dey, A.B.; Pattanayak, S.S.; Mitra, D.; Arif, W. Investigation and design of enhanced decoupled UWB MIMO antenna for wearable applications. *Microw. Opt. Technol. Lett.* **2020**, *63*, 845–861. [CrossRef]
8. Bilal, M.; Saleem, R.; Abbasi, H.H.; Shafique, M.F.; Brown, A.K. An FSS-based nonplanar quad-element UWB-MIMO antenna system. *IEEE Antennas Wirel. Propag. Lett.* **2016**, *16*, 987–990. [CrossRef]
9. Naidu, P.R.T.; Saha, C.; Krishna, K.V.; Shaik, L.A.; Siddiqui, J.Y.; Antar, Y. Compact multiple EBG cells loaded UWB-narrowband antenna pair with high isolation for cognitive radio (CR) based MIMO applications. *AEU-Int. J. Electron. Commun.* **2020**, *127*, 153420. [CrossRef]
10. Kumar, N.; Kiran, K.U. Meander-line electromagnetic bandgap structure for UWB MIMO antenna mutual coupling reduction in E-plane. *AEU-Int. J. Electron. Commun.* **2020**, *127*, 153423. [CrossRef]
11. Modak, S.; Khan, T. A slotted UWB-MIMO antenna with quadruple band-notch characteristics using mushroom EBG structure. *AEU-Int. J. Electron. Commun.* **2021**, *134*, 153673. [CrossRef]

12. Bilal, M.; Shahid, S.; Khan, Y.; Rauf, Z.; Wagan, R.A.; Butt, M.A.; Khonina, S.N.; Kazanskiy, N.L. A Miniaturized FSS-Based Eight-Element MIMO Antenna Array for Off/On-Body WBAN Telemetry Applications. *Electronics* **2022**, *11*, 522. [CrossRef]
13. Amin, F.; Saleem, R.; Shabbir, T.; Rehman, S.U.; Bilal, M.; Shafique, M.F. A compact quad-element UWB-MIMO antenna system with parasitic decoupling mechanism. *Appl. Sci.* **2019**, *9*, 2371. [CrossRef]
14. Kumar, S.; Lee, G.H.; Kim, D.H.; Mohyuddin, W.; Choi, H.C.; Kim, K.W. Multiple-input-multiple-output/diversity antenna with dual band-notched characteristics for ultra-wideband applications. *Microw. Opt. Technol. Lett.* **2020**, *62*, 336–345. [CrossRef]
15. Ramanujam, P.; Venkatesan, P.R.; Arumugam, C.; Ponnusamy, M. Design of miniaturized super wideband printed monopole antenna operating from 0.7 to 18.5 GHz. *AEU-Int. J. Electron. Commun.* **2020**, *123*, 153273. [CrossRef]
16. Kumar, P.; Urooj, S.; Alrowais, F. Design and implementation of quad-port MIMO antenna with dual-band elimination characteristics for ultra-wideband applications. *Appl. Sci.* **2020**, *10*, 1715. [CrossRef]
17. Kumar, P.; Urooj, S.; Malibari, A. Design of quad-port ultra-wideband multiple-input-multiple-output antenna with wide axial-ratio bandwidth. *Sensors* **2020**, *20*, 1174. [CrossRef]
18. Khan, A.A.; Naqvi, S.A.; Khan, M.S.; Ijaz, B. Quad port miniaturized MIMO antenna for UWB 11 GHz and 13 GHz frequency bands. *AEU-Int. J. Electron. Commun.* **2021**, *131*, 153618. [CrossRef]
19. Addepalli, T.; Desai, A.; Elfergani, I.; Anveshkumar, N.; Kulkarni, J.; Zebiri, C.; Rodriguez, J.; Abd-Alhameed, R. 8-Port semi-circular arc MIMO antenna with an inverted L-strip loaded connected ground for UWB applications. *Electronics* **2021**, *10*, 1476. [CrossRef]
20. Sohi, A.K.; Kaur, A. A complementary Sierpinski gasket fractal antenna array integrated with a complementary Archimedean defected ground structure for portable 4G/5G UWB MIMO communication devices. *Microw. Opt. Technol. Lett.* **2020**, *62*, 2595–2605. [CrossRef]
21. Wang, E.; Wang, W.; Tan, X.; Wu, Y.; Gao, J.; Liu, Y. A UWB MIMO slot antenna using defected ground structures for high isolation. *Int. J. RF Microw.-Comput.-Aided Eng.* **2020**, *30*, e22155. [CrossRef]
22. Saadh, A.M.; Ashwath, K.; Ramaswamy, P.; Ali, T.; Anguera, J. A uniquely shaped MIMO antenna on FR4 material to enhance isolation and bandwidth for wireless applications. *AEU-Int. J. Electron. Commun.* **2020**, *123*, 153316. [CrossRef]
23. Anguera, J.; Andújar, A.; Jayasinghe, J.; Chakravarthy, V.; Chowdary, P.; Pijoan, J.L.; Ali, T.; Cattani, C. Fractal antennas: An historical perspective. *Fractal Fract.* **2020**, *4*, 3. [CrossRef]
24. Ray, K.P. Design aspects of printed monopole antennas for ultra-wide band applications. *Int. J. Antennas Propag.* **2008**, *2008*, 713858. [CrossRef]
25. Balanis, C.A. *Antenna Theory: Analysis and Design*; John Wiley & Sons: Hoboken, NJ, USA, 2015.
26. Rafique, U.; Ahmed, M.M.; Hassan, M.M.; Khalil, H. A modified super-wideband planar elliptical monopole antenna. In Proceedings of the 2018 Progress in Electromagnetics Research Symposium (PIERS-Toyama), Toyama, Japan, 1–4 August 2018; pp. 2344–2349.
27. Rafique, U.; Agarwal, S.; Nauman, N.; Khalil, H.; Ullah, K. Inset-fed Planar Antenna Array for Dual-band 5G MIMO Applications. *Prog. Electromagn. Res. C* **2021**, *112*, 83–98. [CrossRef]
28. Agarwal, S.; Rafique, U.; Ullah, R.; Ullah, S.; Khan, S.; Donelli, M. Double Overt-Leaf Shaped CPW-Fed Four Port UWB MIMO Antenna. *Electronics* **2021**, *10*, 3140. [CrossRef]
29. Koohestani, M.; Moreira, A.A.; Skrivervik, A.K. A novel compact CPW-fed polarization diversity ultrawideband antenna. *IEEE Antennas Wirel. Propag. Lett.* **2014**, *13*, 563–566. [CrossRef]

applied sciences

Article

A High-Gain and Wideband MIMO Antenna for 5G mm-Wave-Based IoT Communication Networks

Daniyal Ali Sehrai [1], Muhammad Asif [2,*], Jalal Khan [3], Mujeeb Abdullah [4], Wahab Ali Shah [5], Sattam Alotaibi [6] and Nasim Ullah [6]

1. Electrical Engineering Department, National University of Computer and Emerging Sciences, Islamabad 44000, Pakistan
2. Electrical Engineering Department, University of Science and Technology, Bannu 23200, Pakistan
3. Telecommunication Engineering Department, University of Engineering and Technology, Mardan 23200, Pakistan
4. College of Aeronautical Engineering, National University of Sciences and Technology, Risalpur (Campus), Nowshera 24100, Pakistan
5. Department of Electrical Engineering, Namal University, Mianwali 42250, Pakistan
6. Department of Electrical Engineering, College of Engineering Taif University, Al-Hawiyah, P.O. Box 888, Taif 26571, Saudi Arabia

* Correspondence: masifeed@ustb.edu.pk

Abstract: In this paper, an antenna with a multiple-input, multiple-output (MIMO) configuration is demonstrated for mm-wave 5G-based Internet of Things (IoT) applications. The two antenna elements are arranged next to each other to form a two-port antenna system such that significant field decorrelation is achieved. Moreover, a dielectric layer is backed by an eventual multiport system to amend and analyze the radiation characteristics. The overall size of the MIMO configuration is 14 mm $\times$ 20 mm, and the operation bandwidth achieves ranges from 16.7 to 25.4 GHz, considering the -10 dB criterion with a maximum isolation of more than -30 dB within the operating band. The peak gain offered by the antenna system is nearly 5.48 dB, and incorporating a dielectric layer provides an increase in the gain value to 8.47 dB. Within the operating band, more than 80% total efficiency is observed, and analysis shows several MIMO performance metrics with favorable characteristics. The compactness of the proposed design with high isolation, improved gain, and wideband features make it a suitable candidate for mm-wave-based 5G applications.

Keywords: 5G; mm-wave; IoT; MIMO; communication networks

Citation: Sehrai, D.A.; Asif, M.; Khan, J.; Abdullah, M.; Shah, W.A.; Alotaibi, S.; Ullah, N. A High-Gain and Wideband MIMO Antenna for 5G mm-Wave-Based IoT Communication Networks. *Appl. Sci.* **2022**, *12*, 9530. https://doi.org/10.3390/app12199530

Academic Editor: Christos Bouras

Received: 10 August 2022
Accepted: 19 September 2022
Published: 22 September 2022

Publisher's Note: MDPI stays neutral with regard to jurisdictional claims in published maps and institutional affiliations.

1. Introduction

The increase in annual data use by 40 to 70% has strained capacity requirements, representing an increase by 1000 times with the passage of time. Considering progressing data needs, 5G is a competent candidate adept to offer a data rate of multiple gigabits per second [1–3]. To accomplish the required goals, experts have focused on the mostly unoccupied portion of the spectrum, i.e., mm-wave, where high bandwidth achievement is possible, due in part to the limited bandwidth and capacity constraints of the lower portion of the spectrum (sub 6 GHz) [4,5]. Transmission based on 5G technology is not only intended to have high data rates and high quality of service but also to strengthen abilities of developing technologies, such as the Internet of Vehicles, virtual reality, smart cities etc. [6,7]. Frequency bands for communication networks must be carefully allocated, as several challenges arise with a particular band selection [8], which is why several frequency bands are being considered by various telecom regulatory bodies and various countries. Most of the bands that have been allocated for 5G communication are within the ranges of 24–28 GHz, 37–40 GHz, etc. [9]. However, path loss and atmospheric attenuations factors must be considered and solved before considering transmission at these frequency bands [10]. Furthermore, the bandwidth should be wide enough to offer significantly high data rates [11]. Thus, the role of antennas is key in this regard, and they must be carefully designed to realize 5G communication systems and to effectively benefit users [12].

5G systems will likely be built based on the multiantenna concept on both the transmission and receiver sides, making multiple input, multiple output (MIMO) configurations and also arrays are important in this regard [13]. To tackle attenuation-related issues in the case of mm-wave antennas, arrays are usually utilized to offer a narrow-beam radiation pattern [14]. However, these type of structures are focused on a single feed, and complex and simple power dividers are usually involved [15]. Controlling the losses in these type of power dividers is a challenging task, whereas a single feed results in limited capacity. Accommodating multiple antennas in an array structure to increase the gain also causes an increase in the profile [16]. Likewise, antennas incorporating lenses, metamaterials, and cavities are usually helpful to achieve high gain, and some antennas can maintain a low profile instead of high gain. For example, in [17], a DRA was developed with an EBG placed at a distance of 2.6 mm. The performance of the DRA was evaluated with and without an EBG. Overall, a 3 dB improvement in the gain was achieved. The size of the ground plane was 30 mm × 30 mm for the 60 GHz band. In [18], an EBG ground plane and a dielectric superstrate were utilized to improve the gain of the antenna, resulting in a gain of more than 16 dB. The superstrate was placed 5.35 mm above the main antenna array with a complex power divider. The array structure resulted in a larger PCB size, as a 12 mm separation was used between the antenna elements. A 28 GHz operating frequency antenna was proposed in [19], with a peak gain of 15.6 dB. A fractional bandwidth in the range of 26.5–29.4 GHz was attained. A 3D-printed dielectric lens was utilized to attain high gain. Moreover, 50 mm × 50 mm samples of polylactic acid (PLA) were fabricated. A complex structure was implemented in the proposed method to achieve high gain. However, the MIMO characteristic was lacking in the proposed designs, so capacity issues remained a challenge. Moreover, some of the proposed antennas lack a wide bandwidth. Recently, metamaterials with and without a combination of MIMO have been proposed. Antenna gain was improved by utilizing a three-pair metamaterial arrays in [20]. An overall gain of 7.4 dB was achieved, with bandwidth covering the range from 24.25 to 27.5 GHz and an overall size of 30.5 mm × 30 mm. Another antenna with an MIMO configuration was proposed in [21]; an electromagnetic band gap (EBG) configuration was adopted, increasing the gain by only 1.9 dB. Thus, an overall peak gain of 6 dB was achieved, with a bandwidth as high as 1 GHz. The size of the proposed EBG-based design is 27.5 mm × 27.5 mm. A dielectric resonator (DRA)-based MIMO antenna was proposed in [22], with an operating band range from 26.71 to 28.91 GHz. A peak gain of 7 dB was achieved after incorporating a metasurface with a DRA-based MIMO configuration such that the total size was 20 mm × 40 mm. A metamaterial surface was utilized as a reflector in [23] for an MIMO antenna system (two elements), resulting in a peak gain of 11.5 dB. The overall structure had a larger size of 31.7 mm × 53 mm. In [24], an MIMO antenna system (two elements) with a 8.6 dB peak gain was proposed with an operational bandwidth ranging from 29.7 to 31.5 GHz. Furthermore, the proposed structure had a large size of 48 mm × 21 mm. In [25], a MIMO antenna with a size of 95 mm × 80 mm was presented. The proposed design achieved a wide band but with many rapid jumps in the reflection coefficient, indicating that the reflection coefficient response was not smooth with a stable noise level variation at the transmission frequency band. Moreover, the large size represents a further constraint.

Overall, the use of metamaterial-based surfaces to improve the radiation characteristics of antennas is challenging in terms of controlling the capacitance and inductance in order to achieve a satisfactory output. As an alternative to metamaterial-based reflectors, surfaces are required for which capacitance and inductance can be easily controlled to enhance the radiation characteristics of the antennas, which is desirable for mm-wave communication. Moreover, wide bandwidth with a high gain requirement must be achieved in order to offer a high data rate, which is a promising feature of 5G technology. Due to the limitations observed, in this paper, we present a wideband, high-gain, and multiport antenna system (two antennas with two exciting sources). A compact antenna system is also proposed with a dielectric layer to improve the gain. Furthermore, to efficiently benefit from the MIMO system, high decorrelation in the radiation patterns is a desirable feature. Thus, a wideband ranging from 17.27 to 25.53 GHz with a total efficiency of more than 80% is achieved, with a peak gain of 8.47 dB, an isolation level of more than −30 dB, and the benefit of high decorrelation in radiation patterns.

2. Proposed Antenna Design Procedure

2.1. Antenna Element and MIMO Configuration

The geometry of the proposed design is demonstrated in Figure 1. A Rogers family substrate, i.e., RO-4350B with a thickness of 0.168 mm is used in the proposed design. The proposed system was designed and simulated using computer simulation technology (CST) software. A single element is backed by a truncated ground plane with a front side radiating structure modification to attain a wideband operation with optimal gain, efficiency, etc. Table 1 provides the dimensions summary of the proposed antenna element.

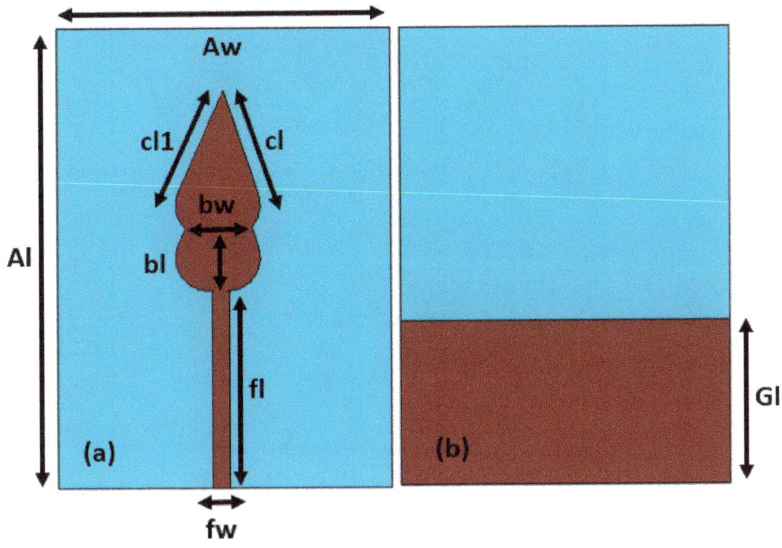

Figure 1. Antenna element geometry: (**a**) front, (**b**) back.

Table 1. Dimensions of the proposed antenna element.

Parameter	Value (mm)	Parameter	Value (mm)
Al	14.0	Aw	10.0
fw	0.5	cl1	3.6
fl	6.0	cl	3.6
bl	1.95	Gl	5.0
bw	2.0	–	–

The following expression can be utilized to estimate the feedline width [14].

$$fw = w_{Z_o} = \left(\frac{377}{Z_o\sqrt{\varepsilon_r}} - 2\right) \times h \quad (1)$$

The Equation (1) is based on a combination of components, i.e., feed line width (*fw*), impedance (Z_o = 50 ohm), substrate thickness (*h*), and dielectric constant (ε_r). The reflection coefficient for the proposed antenna element is analyzed in Figure 2. Based on the −10 dB criterion, the response is well below −10 dB for the desired operational bandwidth. The design evolution steps of the proposed antenna are shown in Figure 2. Initially, a single radiating structure (step 1) provides a narrow bandwidth, whereas introducing a similar radiating structure on the top of the first radiating structure in step 2 yields an improvement in the bandwidth. Finally, in step 3, adding a sharp leaf-type radiating shape on the top of the first two structures results in a wider bandwidth. Thus, the structure achieved in step 3 is selected for subsequent processes, as it operates with a wide bandwidth and a meaningful magnitude of reflection coefficient. The basic antenna element (antenna) is a variation of the well-known circular broadband patch antenna tuned to work in the frequency range of 16.7–25.4 GHz. The mathematical modelling described in [26] was also helpful in determining the final design, including these design evolution steps. The single element is then extended to two-port MIMO configuration, as shown in Figure 3, such that it could be helpful to achieve field decorrelation. The overall dimensions of the proposed MIMO configuration are 14 mm × 20 mm × 0.168 mm.

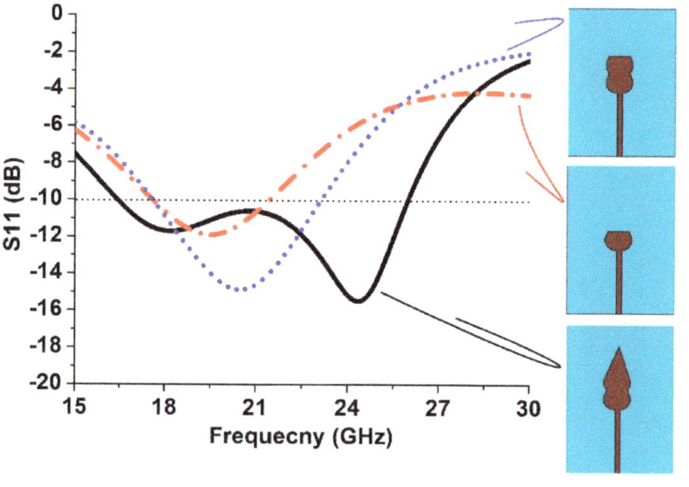

Figure 2. Reflection coefficient comparison of the proposed antenna element design evolution steps.

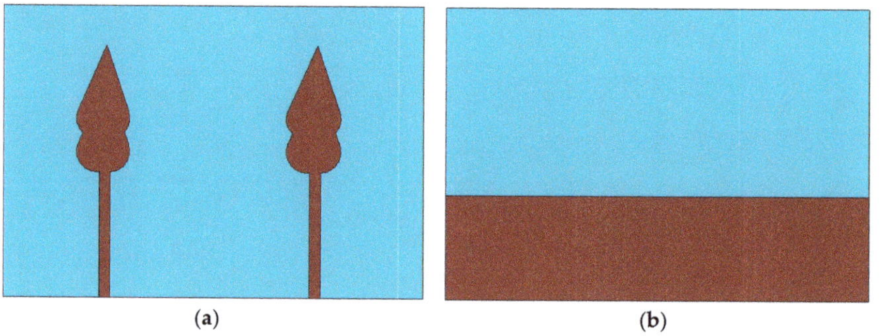

Figure 3. MIMO antenna geometry: (**a**) front, (**b**) back.

The reflection coefficient for the proposed MIMO antenna is shown in Figure 4a, which is well below −10 dB. The isolation between the antenna elements is shown in Figure 4b. It is important to analyze the isolation between the MIMO antenna elements because it becomes an issue when more than one antenna element is assembled on a single printed circuit board (PCB) [27–30]. A maximum isolation level of more than −30 dB is achieved within the operating band. Overall, satisfactory likeness between the measured and simulated results is achieved. In the figures, sim. represents simulated results, whereas mea. represents measured results.

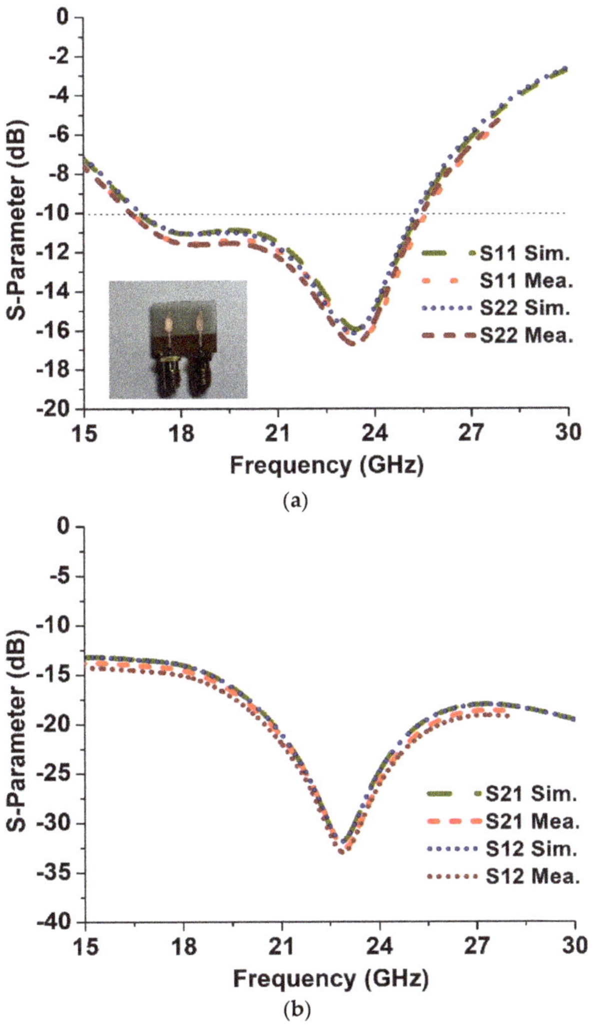

Figure 4. Multiport antenna: (**a**) S-parameters, (**b**) isolation.

2.2. Antenna System Incorporated with a Dielectric Layer

An MIMO antenna incorporated with a dielectric layer is shown in Figure 5. A dielectric layer with overall dimensions of 20 mm × 26 mm × 2 mm is placed below a proposed MIMO configuration at a distance of 4 mm. Moreover, the dielectric constant is maintained at 12 to ensure that the proposed dielectric layer has a significant beam concentration. The overall aim is to improve the radiation performance of the proposed antenna system. The reflection coefficient of the proposed MIMO antenna after incorporating a dielectric layer is analyzed in Figure 6a, and the isolation is shown in Figure 6b, which is more than −30 dB (maximally) within the operating band. Here, we suggest that after incorporating a dielectric layer, both antennas will still achieve similar results, which will be helpful to attain high data rates, which is among the main features of MIMO-based 5G transmission. Furthermore, a gain comparison is shown in Figure 7 for an antenna with and without a dielectric layer. The peak gain in the case of an antenna without an external layer is 5.48 dB, whereas when a dielectric layer is incorporated, the peak gain increases to 8.47 dB within the operating band due to the highly dielectric nature of the layer incorporated on the back of MIMO antenna. Owing to the highly capacitive nature of the dielectric layer, it serves as a suitable reflective surface, resulting in the focusing of the fringing fields and an overall improvement in the gain. In the future, a dielectric layer with nearly the same dielectric constant value will be utilized to validate the simulated results based on the dielectric layer.

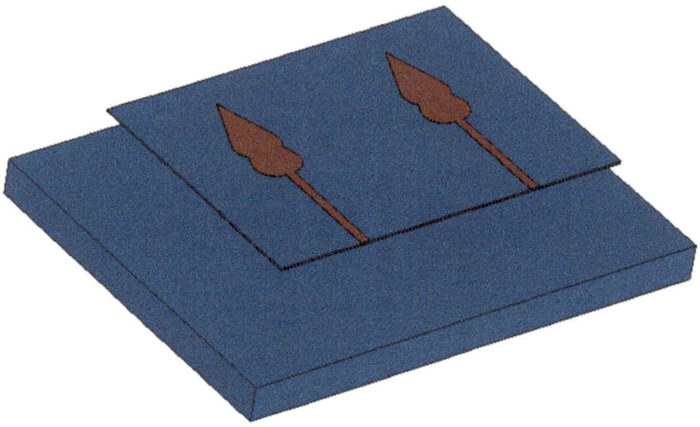

Figure 5. MIMO antenna incorporated with a dielectric layer.

The following expression [31] can be utilized to estimate the distance between the proposed MIMO system and the dielectric layer to achieve a satisfactory output performance.

$$\varphi - 2\beta H = 2n\pi; n = \ldots -1, 0, 1 \tag{2}$$

The equation is based on a combination of components, i.e., free space propagation constant (β), reflection phase (φ), and distance (H) between the dielectric layer and the antenna configuration (π = 3.1415).

2.3. Radiation Pattern in Terms of Gain

In Figure 8, the simulated polar radiation patterns are compared for an antenna with and without a dielectric layer in the E-plane and H-plane in the 23.5 GHz frequency band. To efficiently benefit from the MIMO system, high decorrelation in the radiation patterns is a desirable feature. As shown in Figure 8, ports 1 and 2, the radiation pattern is well-de-correlated, indicating that one-antenna radiation results in low overlapping with the other one antenna within the same panel. This can be help to achieving sufficiently diverse performance. The total efficiency in the desired frequency band without a dielectric layer is 87.8%, whereas after incorporating a dielectric layer, the total efficiency is 89.6%. The gain measured at multiple frequency points for an antenna without a dielectric layer is compared with the simulated gain in Table 2.

Table 2. Gain measurement of the proposed antenna.

Frequency (GHz)	Simulated Value (dB)	Frequency (GHz)	Measured Value (dB)
21.5	4.61	21.5	4.45
22.5	5.07	22.5	4.85
23.5	5.48	23.5	5.34

2.4. MIMO Performance Metrics

Several MIMO performance metrics were analyzed for the proposed antenna system incorporated with a dielectric layer, such as envelope correlation coefficient (ECC), channel capacity loss (CCL), and diversity gain (DG). In Figure 9, the ECC is demonstrated using the far-field calculation method [31], which conforms with the standard value (<0.5), whereas in Figure 10, DG is analyzed, which is close to the standard value, i.e., 10 dB. Moreover, in Figure 11, CCL is shown. A CCL value of less than 0.4 bits/s/Hz is achieved across the operating band, which depicts a minimal capacity loss. It is important to analyze this loss because when more than one antenna is incorporated to form an antenna system, the channel capacity is increased, but some losses are encountered due to the correlation between the MIMO links. In [32], a channel capacity of 41 bps/Hz was observed for the sub 6 GHz band, with a low isolation value of −14.8 dB. CCL was not analyzed, and the size of the proposed design was relatively large, i.e., >80 mm. An MIMO antenna with a size of 150 mm × 80 mm was presented in [33]; a channel capacity of 41 bps/Hz was achieved, but no analysis of CCL was presented. The proposed MIMO antenna achieved an isolation of less than −15 dB within the 5150–5925 MHz frequency band, which may result in deterioration of the transmission quality. The same idea was implemented in [34], in which CCL was not taken into account and a relatively larger antenna of more than 80 mm was reported to have a low operating bandwidth, i.e., less than 1 GHz.

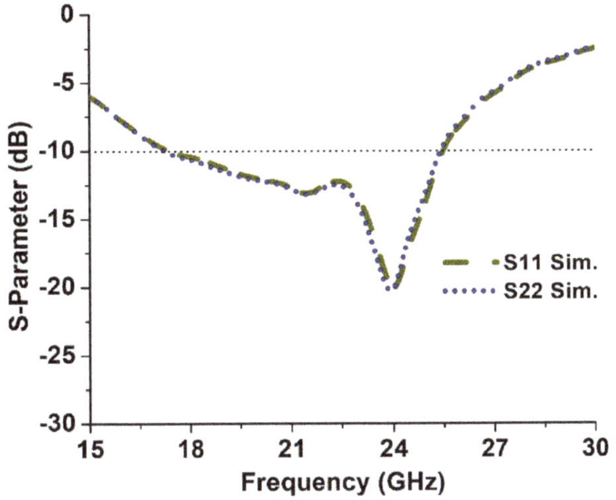

(a)

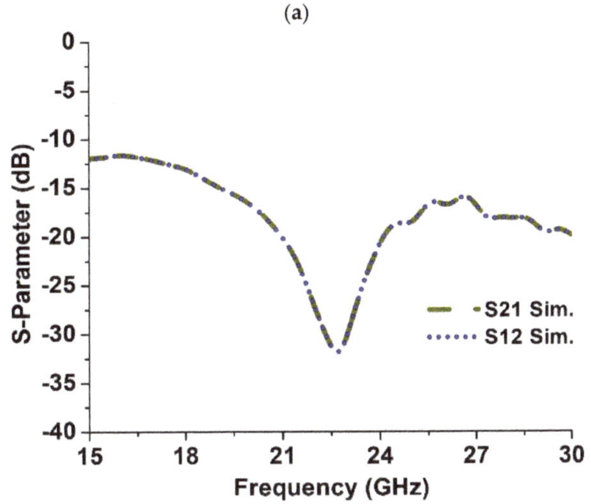

(b)

Figure 6. S-parameters based on antenna incorporation with a dielectric layer: (**a**) reflection coefficient, (**b**) isolation.

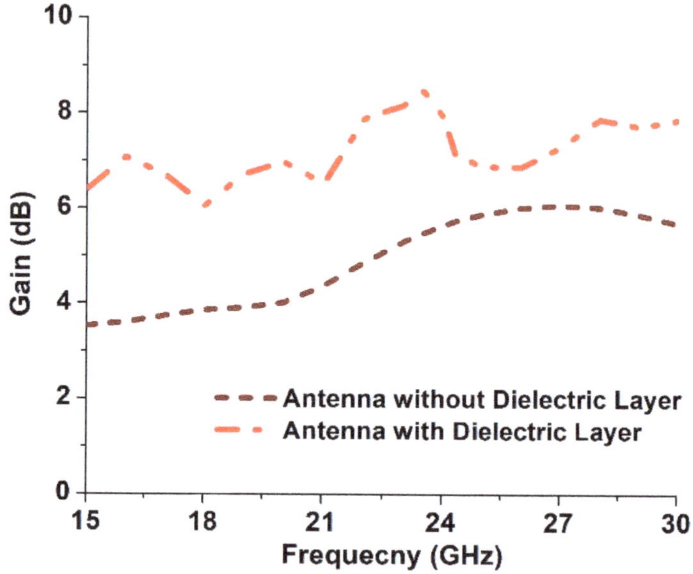

Figure 7. Proposed MIMO antenna gain comparison with and without a dielectric layer.

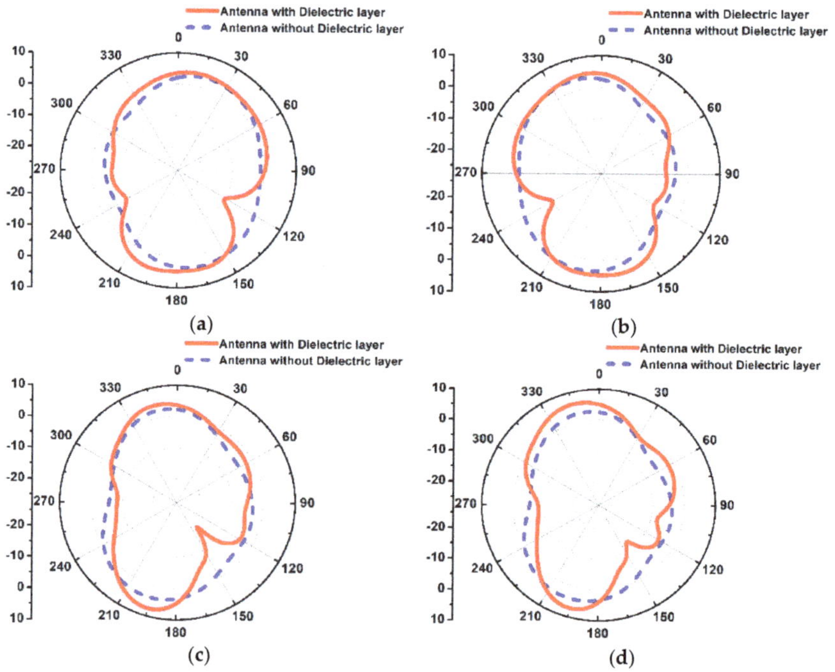

Figure 8. Antenna polar radiation patterns in: (**a**) E-plane port 1, (**b**) E-plane port 2, (**c**), H-plane port 1, and (**d**) H-plane port 2.

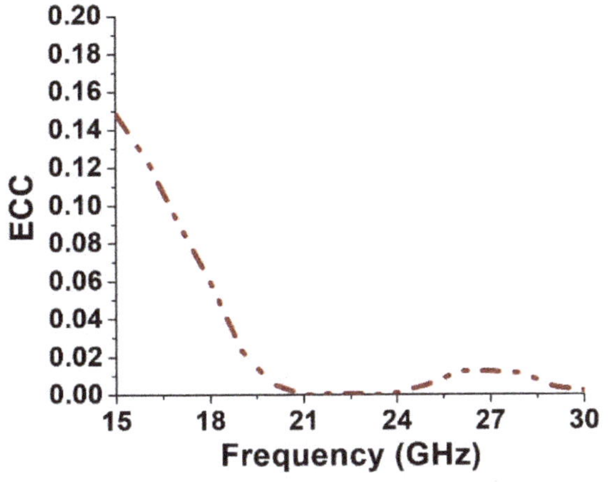

Figure 9. ECC of the proposed antenna system incorporated with a dielectric layer.

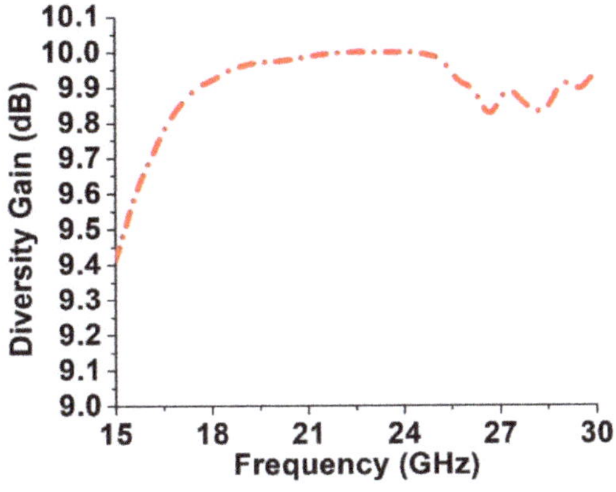

Figure 10. DG of the proposed antenna system incorporated with a dielectric layer.

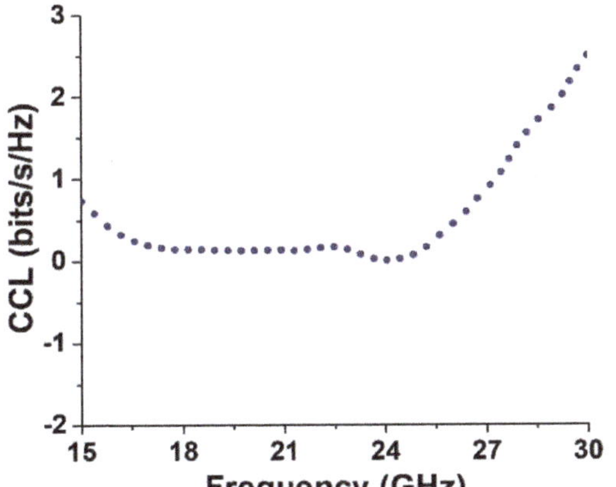

Figure 11. CCL of the proposed antenna system incorporated with a dielectric layer.

3. Conclusions

In this paper, we demonstrated an antenna with a multiple input, multiple output (MIMO) configuration for mm-wave 5G applications. The two antenna elements are sequenced in series to form a two-port antenna system such that significant field decorrelation is achieved. Moreover, a dielectric layer is backed by the proposed MIMO antenna to improve and analyze the radiation characteristics. The overall size of the MIMO configuration is 14 mm $\times$ 20 mm, with an operation bandwidth in the range of 16.7–25.4 GHz, considering the -10 dB criteria with a maximum isolation of more than -30 dB within the operating band. The peak gain offered by the proposed antenna system is nearly 5.48 dB, and deploying a dielectric layer increases the gain value to 8.47 dB. A total efficiency of more than 80% is achieved within the operating band, and several MIMO performance metrics were analyzed, revealing satisfactory characteristics, making the proposed design a suitable candidate for mm-wave-based 5G applications. In the future, beam-steering techniques will be incorporated to achieve satisfactory cell transmission coverage.

Author Contributions: Conceptualization, D.A.S.; Data curation, M.A. (Mujeeb Abdullah); Formal analysis, W.A.S.; Funding acquisition, N.U.; Investigation, J.K.; Project administration, S.A.; Supervision, M.A. (Muhammad Asif). All authors have read and agreed to the published version of the manuscript.

Funding: The authors would like to thank Taif University Researchers, Taif University, Taif, Saudi Arabia, for supporting project number TURSP-2020/228.

Conflicts of Interest: The authors declare that they have no conflict of interest to report regarding the present study.

References

1. Kiani, S.H.; Iqbal, A.; Wong, S.-W.; Savci, H.S.; Alibakhshikenari, M.; Dalarsson, M. Multiple elements MIMO antenna system with broadband operation for 5th generation smart phones. *IEEE Access* **2022**, *10*, 38446–38457. [CrossRef]
2. Alibakhshikenari, M.; Ali, E.M.; Soruri, M.; Dalarsson, M.; Naser-Moghadasi, M.; Virdee, B.S.; Stefanovic, C.; Pietrenko-Dabrowska, A.; Koziel, S.; Szczepanski, S.; et al. A comprehensive survey on antennas on-chip based on metamaterial, metasurface, and substrate integrated waveguide principles for millimeter-waves and terahertz integrated circuits and systems. *IEEE Access* **2022**, *10*, 3668–3692. [CrossRef]
3. Thompson, J.; Ge, X.; Wu, H.; Irmer, R.; Jiang, H.; Fettweis, G.; Alamouti, S. 5G wireless communication systems: Prospects and challenges [guest editorial]. *IEEE Commun. Mag.* **2014**, *52*, 62–64. [CrossRef]
4. Wei, L.; Hu, R.Q.; Qian, Y.; Wu, G. Key elements to enable millimeter wave communications for 5G wireless systems. *IEEE Wireless Commun.* **2014**, *21*, 136–143.
5. Hussain, M.; Mousa Ali, E.; Jarchavi, S.M.R.; Zaidi, A.; Najam, A.I.; Alotaibi, A.A.; Althobaiti, A.; Ghoneim, S.S.M. Design and characterization of compact broadband antenna and its MIMO configuration for 28 GHz 5G applications. *Electronics* **2022**, *11*, 523. [CrossRef]
6. Rappaport, T.S.; Sun, S.; Mayzus, R.; Zhao, H.; Azar, Y.; Wang, K.; Wong, G.N.; Schulz, J.K.; Samimi, M.; Gutierrez, F. Millimeter wave mobile communications for 5G cellular: It will work! *IEEE Access* **2013**, *1*, 335–349. [CrossRef]
7. Alibakhshikenari, M.; Virdee, B.S.; See, C.H.; Shukla, P.; Moghaddam, S.M.; Zaman, A.U.; Shafqaat, S.; Akinsolu, M.O.; Liu, B.; Yang, J.; et al. Dual-polarized highly folded bowtie antenna with slotted self-grounded structure for Sub-6 GHz 5G applications. *IEEE Trans. Antennas Propag.* **2022**, *70*, 3028–3033. [CrossRef]
8. European 5G Observatory. National 5G Spectrum Assignment. Available online: https://5gobservatory.eu/ (accessed on 10 May 2020).
9. Hussain, N.; Jeong, M.; Abbas, A.; Kim, N. Metasurface-based single-layer wideband circularly polarized MIMO antenna for 5G millimeter-wave systems. *IEEE Access* **2020**, *8*, 130293–130304. [CrossRef]
10. Ojaroudiparchin, N.; Shen, M.; Pedersen, G.F. Beam-steerable microstrip-fed bow-tie antenna array for fifth generation cellular communications. In Proceedings of the 2016 10th European Conference on Antennas and Propagation (EuCAP), Davos, Switzerland, 10–15 April 2016; pp. 1–5.
11. Sharaf, M.H.; Zaki, A.I.; Hamad, R.K.; Omar, M.M.M. A Novel Dual-Band (38/60 GHz) Patch Antenna for 5G Mobile Handsets. *Sensors* **2020**, *20*, 2541. [CrossRef]
12. Barreto, A.N.; Faria, B.; Almeida, E.; Rodriguez, I.; Lauridsen, M.; Amorim, R.; Vieira, R. 5G–Wireless Communications for 2020. *J. Commun. Inf. Syst.* **2016**, *31*. [CrossRef]
13. Li, M.Y.; Ban, Y.L.; Xu, Z.Q.; Guo, J.; Yu, Z.F. Tri-polarized 12-Antenna MIMO array for future 5G smartphone applications. *IEEE Access* **2017**, *6*, 6160–6170. [CrossRef]

14. Khan, J.; Sehrai, D.A.; Ali, U. Design of dual band 5G antenna array with SAR analysis for future mobile handsets. *J. Electr. Eng. Technol.* **2019**, *14*, 809–816. [CrossRef]
15. Li, Y.; Luk, K.-M. Wideband perforated dense dielectric patch antenna array for millimeter-wave applications. *IEEE Trans. Antennas Propag.* **2015**, *63*, 3780–3786. [CrossRef]
16. Ta, S.X.; Choo, H.; Park, I. Broadband printed-dipole antenna and its arrays for 5G applications. *IEEE Antennas Wireless Propag. Lett.* **2017**, *16*, 2183–2186. [CrossRef]
17. Al-Hasan, M.J.; Denidni, T.A.; Sebak, A.R. Millimeter-wave EBG-based aperture-coupled dielectric resonator antenna. *IEEE Trans. Antennas Propag.* **2013**, *61*, 4354–4357. [CrossRef]
18. Haraz, O.M.; Elboushi, A.; Alshebeili, S.A.; Sebak, A. Dense dielectric patch array antenna with improved radiation characteristics using EBG ground structure and dielectric superstrate for future 5G cellular networks. *IEEE Access* **2014**, *2*, 909–913. [CrossRef]
19. Malik, B.T.; Doychinov, V.; Zaidi, S.A.R.; Robertson, I.D.; Somjit, N. Antenna gain enhancement by using low-infill 3D-printed dielectric lens antennas. *IEEE Access* **2019**, *7*, 102467–102476. [CrossRef]
20. Attia, H.; Abdelghani, M.L.; Denidni, T.A. Wideband and high-gain millimeter-wave antenna based on FSS Fabry–Perot cavity. *IEEE Trans. Antennas Propag.* **2017**, *65*, 5589–5594. [CrossRef]
21. Iqbal, A.; Basir, A.; Smida, A.; Mallat, N.K.; Elfergani, I.; Rodriguez, J.; Kim, S. Electromagnetic bandgap backed millimeter-wave MIMO antenna for wearable applications. *IEEE Access* **2019**, *7*, 111135–111144. [CrossRef]
22. Murthy, N.S. Improved isolation metamaterial inspired mm-wave MIMO dielectric resonator antenna for 5G application. *Prog. Electromagn. Res. C* **2020**, *100*, 247–261. [CrossRef]
23. Saad, A.A.R.; Mohamed, H.A. Printed millimeter-wave MIMO based slot antenna arrays for 5G networks. *AEU Int. J. Electron. Commun.* **2019**, *99*, 59–69. [CrossRef]
24. Sharawi, M.S.; Podilchak, S.K.; Hussain, M.T.; Antar, Y.M.M. Dielectric resonator based MIMO antenna system enabling millimeter wave mobile devices. *IET Microw. Antennas Propag.* **2017**, *11*, 287–293. [CrossRef]
25. Gotra, S.; Varshney, G.; Pandey, V.S.; Yaduvanshi, R.S. Super-wideband multi-input–multi-output dielectric resonator antenna. *IET Microw. Antennas Propag.* **2019**, *14*, 21–27. [CrossRef]
26. Ahmad, S.; Ijaz, U.; Naseer, S.; Ghaffar, A.; Qasim, M.A.; Abrar, F.; Parchin, N.O.; See, C.H.; Abd-Alhameed, R. A jug-shaped CPW-fed ultra-wideband printed monopole antenna for wireless communications networks. *Appl. Sci.* **2022**, *12*, 821. [CrossRef]
27. Alibakhshikenari, M.; Khalily, M.; Virdee, B.S.; See, C.H.; Abd-Alhameed, R.A.; Limiti, E. Mutual-coupling isolation using embedded metamaterial EM bandgap decoupling slab for densely packed array antennas. *IEEE Access* **2019**, *7*, 51827–51840. [CrossRef]
28. Alibakhshikenari, M.; Babaeian, F.; Virdee, B.S.; Aïssa, A.; Azpilicueta, L.; See, C.H.; Althuwayb, A.A.; Huynen, I.; Abd-Alhameed, R.A.; Falcone, F.; et al. A comprehensive survey on "various decoupling mechanisms with focus on metamaterial and metasurface principles applicable to SAR and MIMO antenna systems". *IEEE Access* **2020**, *8*, 192965–193004. [CrossRef]
29. Alibakhshikenari, M.; Khalily, M.; Virdee, B.S.; See, C.H.; Abd-Alhameed, R.A.; Limiti, E. Mutual coupling suppression between two closely placed microstrip patches using EM-bandgap metamaterial fractal loading. *IEEE Access* **2019**, *7*, 23606–23614. [CrossRef]
30. Alibakhshikenari, M.; Virdee, B.S.; See, C.H.; Abd-Alhameed, R.A.; Ali, A.H.; Falcone, F.; Limiti, E. Study on isolation improvement between closely-packed patch antenna arrays based on fractal metamaterial electromagnetic bandgap structures. *IET Microw. Antennas Propag.* **2018**, *12*, 2241–2247. [CrossRef]
31. Asif, M.; Sehrai, D.A.; Kiani, S.H.; Khan, J.; Abdullah, M.; Ibrar, M.; Alibakhshikenari, M.; Falcone, F.; Limiti, E. Design of a Dual Band SNG Metamaterial Based Antenna for LTE 46/WLAN and Ka-Band Applications. *IEEE Access* **2021**, *9*, 71553–71562. [CrossRef]
32. Thakur, V.; Jaglan, N.; Gupta, S.D. Side edge printed eight-element compact MIMO antenna array for 5G smartphone applications. *J. Electromagn. Waves Appl.* **2022**, *36*, 1685–1701. [CrossRef]
33. Jaglan, N.; Gupta, S.D.; Kanaujia, B.K.; Sharawi, M.S. 10 element sub-6-GHz multi-band double-T based MIMO antenna system for 5G smartphones. *IEEE Access* **2021**, *9*, 118662–118672. [CrossRef]
34. Jaglan, N.; Gupta, S.D.; Sharawi, M.S. 18 Element Massive MIMO/Diversity 5G Smartphones Antenna Design for Sub-6 GHz LTE Bands 42/43 Applications. *IEEE Open J. Antennas Propag.* **2021**, *2*, 533–545. [CrossRef]

applied sciences

MDPI

Article

An Elliptical-Shaped Dual-Band UWB Notch Antenna for Wireless Applications

Om Prakash Kumar [1], Tanweer Ali [1,*], Pramod Kumar [1], Pradeep Kumar [2] and Jaume Anguera [3,4,*]

1 Department of Electronics and Communication, Manipal Institute of Technology, Manipal Academy of Higher Education, Manipal 576104, India
2 Discipline of Electrical, Electronic and Computer Engineering, University of KwaZulu-Natal, Durban 4041, South Africa
3 Fractus Antennas, 08174 Barcelona, Spain
4 Telecommunication Engineering, Universitat Ramon Llull, 08022 Barcelona, Spain
* Correspondence: tanweer.ali@manipal.edu (T.A.); jaume.anguera@salle.url.edu (J.A.)

Abstract: This paper discusses an elliptical ultrawideband (UWB) antenna and a dual-band UWB notch antenna. To achieve a UWB bandwidth, two corner cuts are etched into the rectangular slot on the partial ground plane. An inverted-U-shaped and conductor-shaped resonator are utilized to achieve dual-band notch characteristics on a partial ground plane. The suggested antenna has an overall dimension of 24×32 mm^2. The suggested UWB antenna has a gain of 4.9 dB, a bandwidth of 2.5–11 GHz, a linear phase response, a group delay of less than 1 ns, and a steady radiation pattern. The suggested UWB notch rejects WLAN and ITU bands from 5.2–5.7 GHz and 7.2–8.5 GHz, respectively, with an impedance bandwidth of 2.5–11 GHz. The UWB notch antenna features a linear phase, a group delay of less than 1 ns, and a stable radiation pattern.

Keywords: ultrawideband (UWB); UWB antenna; parasitic resonator; dual-band-notched; monopole antenna

Citation: Kumar, O.P.; Ali, T.; Kumar, P.; Kumar, P.; Anguera, J. An Elliptical-Shaped Dual-Band UWB Notch Antenna for Wireless Applications. *Appl. Sci.* **2023**, *13*, 1310. https://doi.org/10.3390/app13031310

Academic Editors: Dimitris Mourtzis, Daniele Funaro and Christos Bouras

Received: 15 November 2022
Revised: 28 December 2022
Accepted: 12 January 2023
Published: 18 January 2023

1. Introduction

A UWB bandwidth is the frequency band bounded by the points that are 10 dB below the highest radiated emission, as based on the complete transmission system including the antenna. The upper boundary is designated f_H, and the lower boundary is designated f_L. An antenna having a fractional bandwidth equal to or greater than 0.20 or having a UWB bandwidth equal to or greater than 500 MHz, regardless of the fractional bandwidth, is called a UWB antenna. [1,2]. UWB communication, as the name implies, has a large bandwidth and supports a high transmission rate. Ultrawideband antenna applications have recently expanded to cover many computing devices and IoT peripherals [3]. The bulk of these wireless communication systems make use of a UWB antenna for purposes including but not limited to data transfer, reception, localization, location identification, sensing, and tracking.

UWB antenna design varies depending on the communication system's radiation characteristics, and each application necessitates specialized UWB antennas. Due to developments in impulse technology, low-power communication systems, and the availability of unlicensed frequency bands, UWB has emerged as the dominant technique for short-range wireless communication [4]. Due to its inherent properties, UWB technology is well suited for precisely obtaining spatial and directional data, including a high data rate, a broad bandwidth, a reduction in multipath fading, a low cost, and low power consumption [5].

The recent proliferation of wireless communication, wireless portable devices, and smart consumer electronics has increased demand for ultrawideband antenna applications [6]. UWB antenna design research [7–10] is undertaken to improve UWB communication systems with new breakthroughs. Moreover, a single UWB antenna may span numerous frequencies, transmission functions, and operation bands. Therefore, their utilization can eliminate the need for many narrowband antennas. The interference caused by several antennas is decreased, and antenna space is conserved. It is crucial to remember that to get the most out of UWB antennas, they must be carefully designed for the individual needs of each system.

Narrowband systems such as WiMAX, WLAN, and C-band, among others, cause significant interference to UWB antennas. Strong disruptions, such as background noise, emerge from these interferences [11]. A notch is built in the UWB antenna to eliminate these intense narrowband interferences. Notch is the technique of removing unwanted frequency bands from the UWB spectrum [12].

In the literature, many UWB notch antennas have been recorded. The authors of article [13] suggest a floret antenna for wideband application. A UWB antenna with an arch-shaped ground plane is proposed using Koch fractal on the hexagonal pattern patch [14]. In paper [15], a UWB antenna with a slot-type split-ring resonator and a defective ground plane is proposed to achieve a dual notch. A dual-band notch antenna using sectorial and inverted U-shaped slots over a rectangular patch is reported in the article [16]. Paper [17] presents a fractal dual-band UWB antenna with a C-shaped complementary split-ring resonator on the patch's transmission line to perform dual-notch filtering. A dual-notch band UWB antenna is proposed in the article [18] using C-slot and inverted U-slot. A UWB antenna in [19] uses a split-ring slot to create a dual notch. A wide bandwidth in the UWB range is achieved by altering the rectangular patch. In [20], the authors suggest a notch antenna with two bands. A meandering patch slot and a U-shaped feed slot provide a 3–3.9 and 5.22–5.7 GHz notch band. In [21], the patch's two U-shaped stubs create a 3.0–3.9 GHz and 5.0–5.9 GHz notch band.

In this article, an elliptical UWB notch antenna has been designed and fabricated to cover UWB channel 1 to 15, i.e., low band (UWB channels 1 to 4) and high band (UWB channels 5 to 15). The addition of parasitic resonators for UWB notch antenna overcomes electromagnetic interference between bands by inserting inverted-U-shaped and iron-shaped parasitic resonators on the ground plane. Evolution steps from the UWB antenna to the UWB notch antenna are discussed. Sections 2 and 3 discuss the technique for designing antennas. Section 4 discusses result analysis. Section 5 discusses the conclusion followed by references.

2. UWB Antenna Design Approach

The evolution of the UWB antenna is depicted in Figure 1, while the S_{11} characteristics are depicted in Figure 2. In the first stage, an ellipse radiating patch and a rectangle ground plane are created. However, the S_{11} value is much higher than -10 dB.

As a result, the ground plane was further modified in stage 2 to become a partly rectangular ground plane. The S_{11} value decreases as a result. In stage 3, a partially rectangular ground plane is slotted to improve S_{11}. The rectangular slot on the ground plane significantly brings the S_{11} value below -10 dB, but it does not cover the UWB range. As a result, in stage 4 of Figure 1, more modifications are made to obtain UWB characteristics in the antenna. Two corner cuts are etched on the rectangular slot on the ground plane. It gives the impedance bandwidth from 2.5–11 GHz, which covers the UWB range.

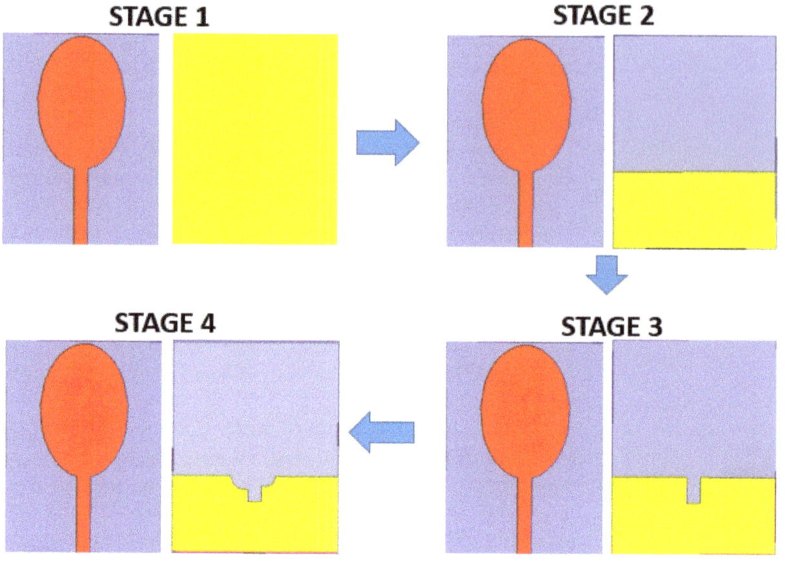

Figure 1. Progression stage of UWB antenna.

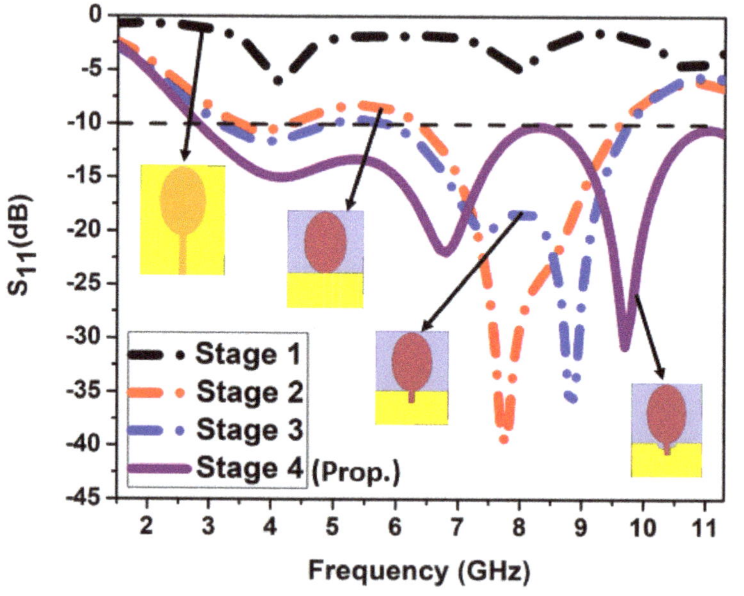

Figure 2. S_{11} characteristics of UWB antenna.

3. UWB Notch Antenna Design Approach

Figure 3 shows the design of an elliptical UWB notch antenna. After completing the design of the elliptical-shaped UWB antenna, the notches have to be introduced. There are many techniques to obtain notches. Two parasitic resonators, an inverted-U-shaped resonator and an iron-shaped resonator, are used to achieve the notch. The inverted-U-shaped resonator helps obtain a notch at 5.3 GHz, and the iron-shaped resonator helps in obtaining a notch at 8.1 GHz. Figure 4 shows the dimension of the suggested antenna. Table 1 lists the antenna's dimensions.

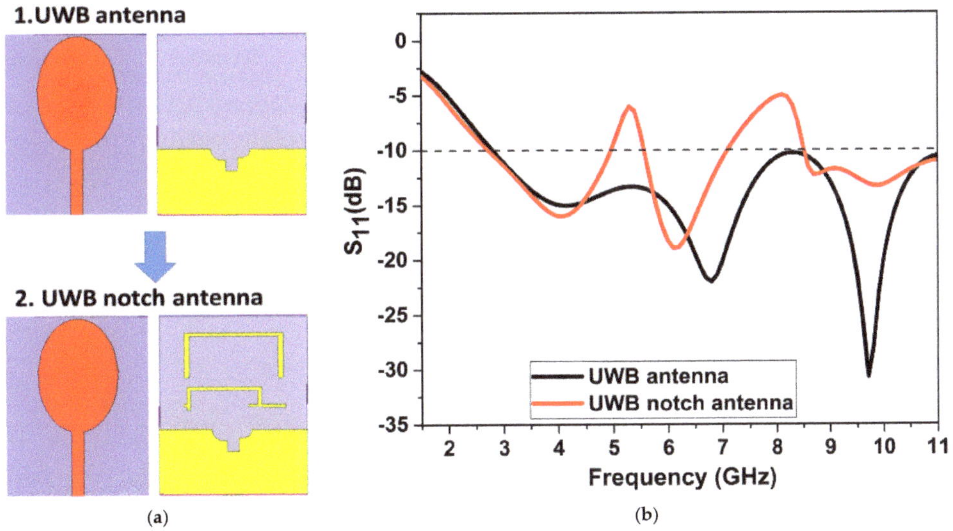

Figure 3. Elliptical UWB notch antenna: (**a**) UWB to UWB notch; (**b**) S_{11}.

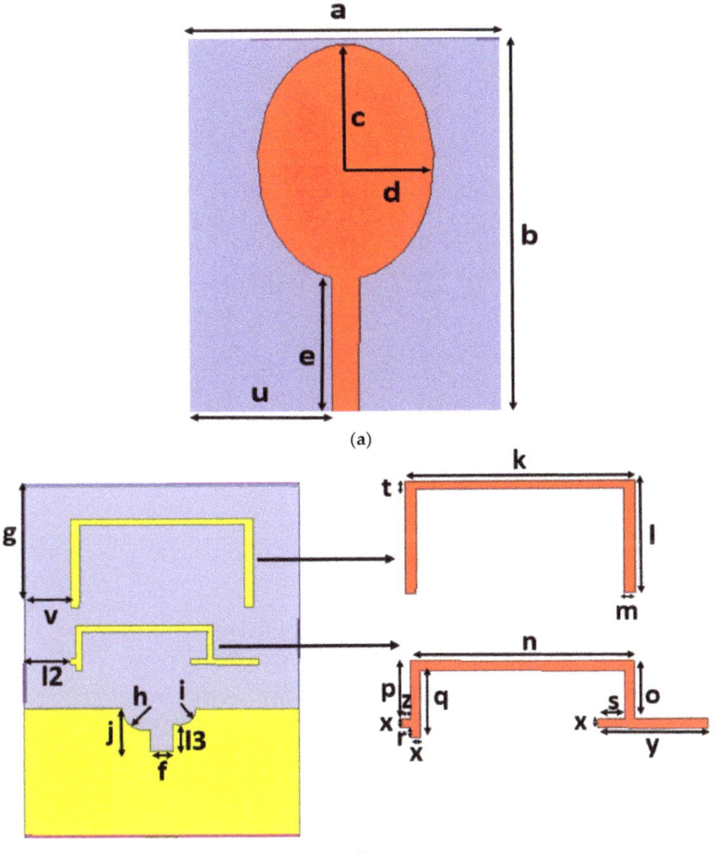

Figure 4. Dimension of elliptical UWB notch antenna: (**a**) patch; (**b**) ground plane.

Table 1. Dimension of the proposed antenna.

Parameter	Dimension (mm)	Parameter	Dimension (mm)
a	24	i	1.25
b	32	f	2
c	9.5	t	0.5
d	6	k	16
e	11.6	l	8
u	11	m	0.75
g	11.25	m	0.75
v	4	n	12
$l2$	4	o	3
$l3$	2.47	p	3
j	1.9	q	3.5
h	1.5	r	0.5
s	1.5	x	0.5
y	6	z	0.5

The length of truncation in the ground plane for achieving UWB is calculated using Equation (1).

$$\begin{aligned} L_{trunc} &= h(perimeter) + j + f + l3 + i(perimeter) \\ &= 1.5 + 1.9 + 2 + 2.47 + 1.25 \\ &= 9.12 \text{ mm} \end{aligned} \tag{1}$$

The relative permittivity of the FR-4 substrate is $\varepsilon_r = 4.4$, so effective permittivity can be calculated as Equation (2),

$$\begin{aligned} \varepsilon_{eff} &= \tfrac{\varepsilon_r+1}{2} + \tfrac{\varepsilon_r-1}{2}\left[1 + 12\tfrac{H}{w}\right]^{-0.5} \\ &= 2.7 + \frac{1.7}{\sqrt{1+12\left(\frac{1.6\times10^{-3}}{3\times10^{-3}}\right)}} = 3.325 \end{aligned} \tag{2}$$

As we can observe that the first notch frequency, f_{notch1} is around 5.2 GHz, and is due to the effect of the upper resonator. Therefore, mathematically, the length of the upper resonator can be calculated using Equation (3).

$$\begin{aligned} L_{upper\ resonator} &= \frac{\lambda_0}{2\sqrt{\varepsilon_{reff}}} = \frac{c}{2\, f_{notch1}\sqrt{\varepsilon_{reff}}} \\ &= \frac{3\times10^{-8}}{2\times5.2\times10^{9}\sqrt{3.325}} = 15.55 \text{ mm} \end{aligned} \tag{3}$$

Similarly, the second notch frequency, f_{notch2} is around 8.2 GHz and is due to the effect of the lower resonator. Therefore, mathematically, the length of the lower resonator can be calculated using Equation (4).

$$\begin{aligned} L_{lower\ resonator} &= \frac{\lambda_0}{2\sqrt{\varepsilon_{reff}}} = \frac{c}{2\, f_{notch2}\sqrt{\varepsilon_{reff}}} \\ &= \frac{3\times10^{-8}}{2\times8.2\times10^{9}\sqrt{3.325}} = 10 \text{ mm} \end{aligned} \tag{4}$$

3.1. *Progression of UWB Notch Antenna*

This section describes the evolution stages of a UWB notch antenna. It shows a step-by-step iterative process to arrive at the final structure. The following Figure 5 shows a pictorial representation of the iterative process.

A patch and a ground plane make up the proposed antenna. Both of these have undergone modification to arrive at the final structure. The S_{11} characteristics of the UWB notch antenna at every stage are plotted together and depicted in Figure 6.

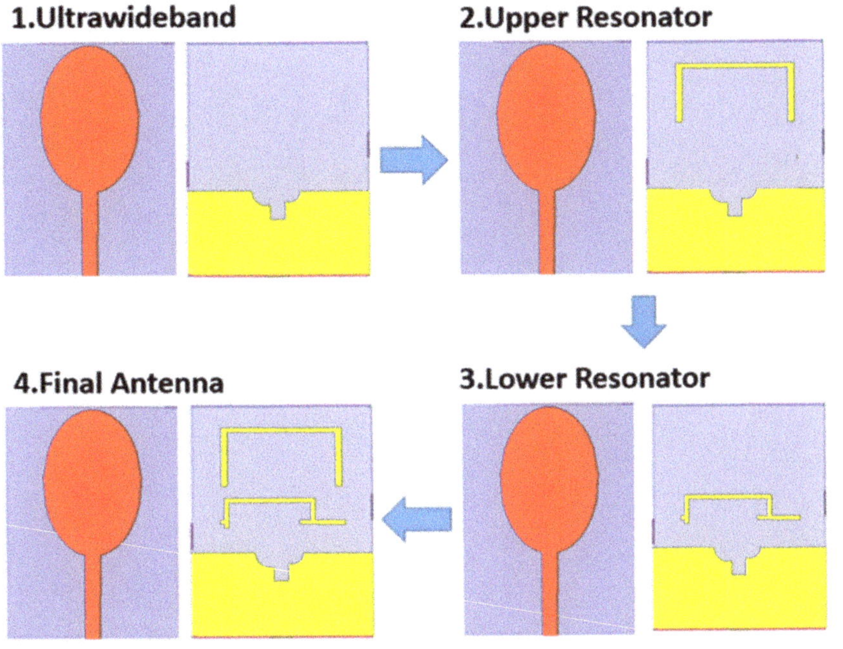

Figure 5. Progression stage of UWB notch antenna.

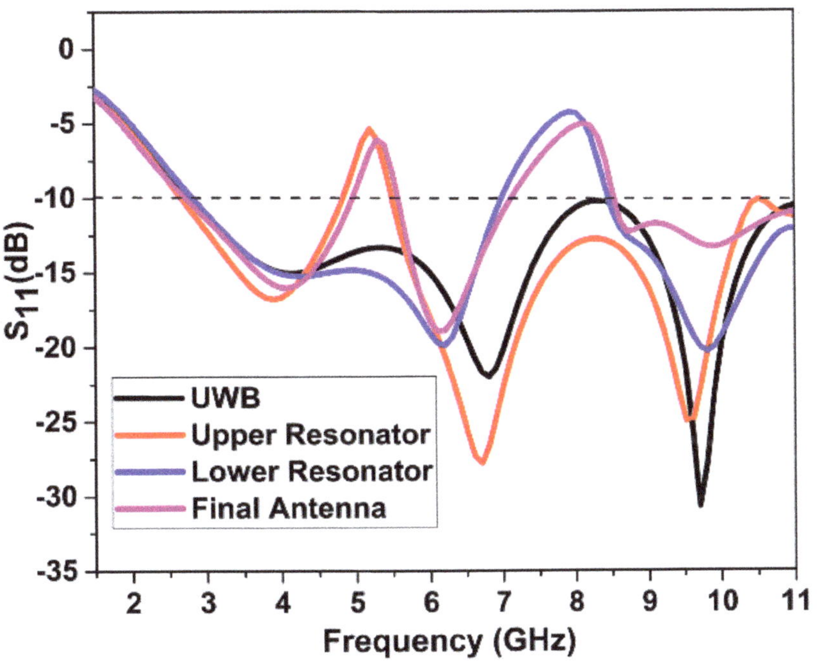

Figure 6. S11 characteristics of progression stage of UWB notch antenna.

In step 1, the UWB antenna is designed. The ground plane is then etched with an in-verted-U-shaped upper resonator in step 2. The inverted-U-shaped upper resonator gives the first notch at 5.3 GHz, as shown in Figure 5. In step 3, an iron-shaped lower resonator is added to the ground plane. The iron-shaped lower resonator gives the sec notch at 8.1 GHz. Thus, the upper resonator and lower resonator combination make elliptical UWB notch antenna in step 4.

These graphs in Figure 6 show that the notch antenna has $S_{11} < -10$ dB for the entir UWB range and has two notches at 5.3 GHz and 8.1 GHz due to the two resonators.

3.2. Parametric Analysis of UWB Notch Antenna

The impact of the stubs dimension on the notch antenna is examined using a parametric analysis. The dimensions t and x have an impact on the design's performance.

3.2.1. Effect of Upper Resonator-Width 't'

To analyze the effects of width 't' on the S_{11} characteristics, the variation in width 't' was simulated and analyzed. The S_{11} characteristics are represented in Figure 7. When the width is 0.5 mm, the result is optimum.

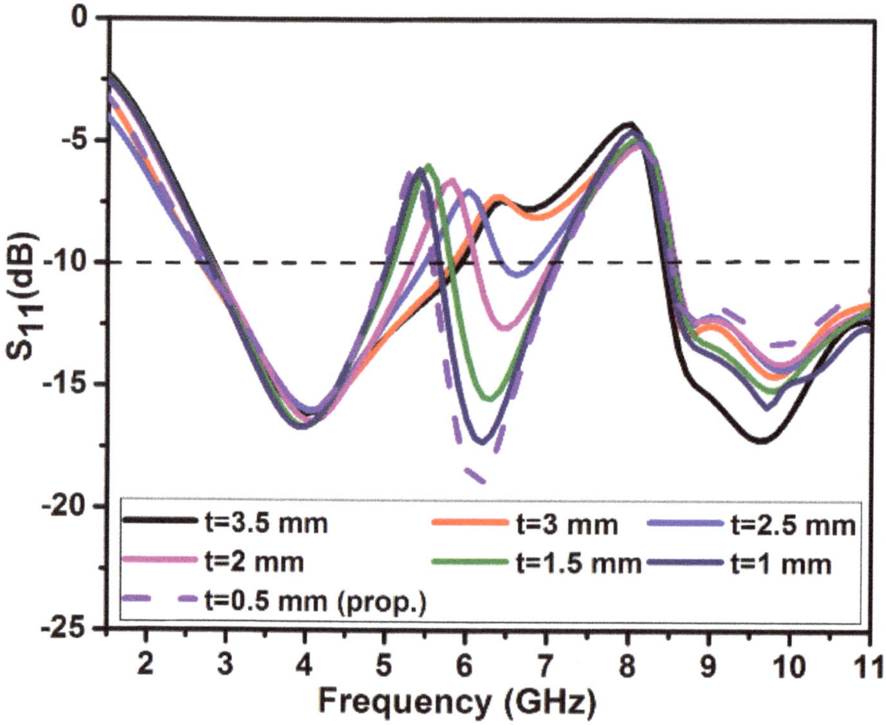

Figure 7. S_{11} for variation of width 't'.

3.2.2. Effect of Lower Resonator-Width 'x'

The variation in width 'x' is simulated and studied to investigate the effects of width 'x' on the S_{11} features. Figure 8 illustrates the S_{11} features. The result is best when the width is 0.5 mm.

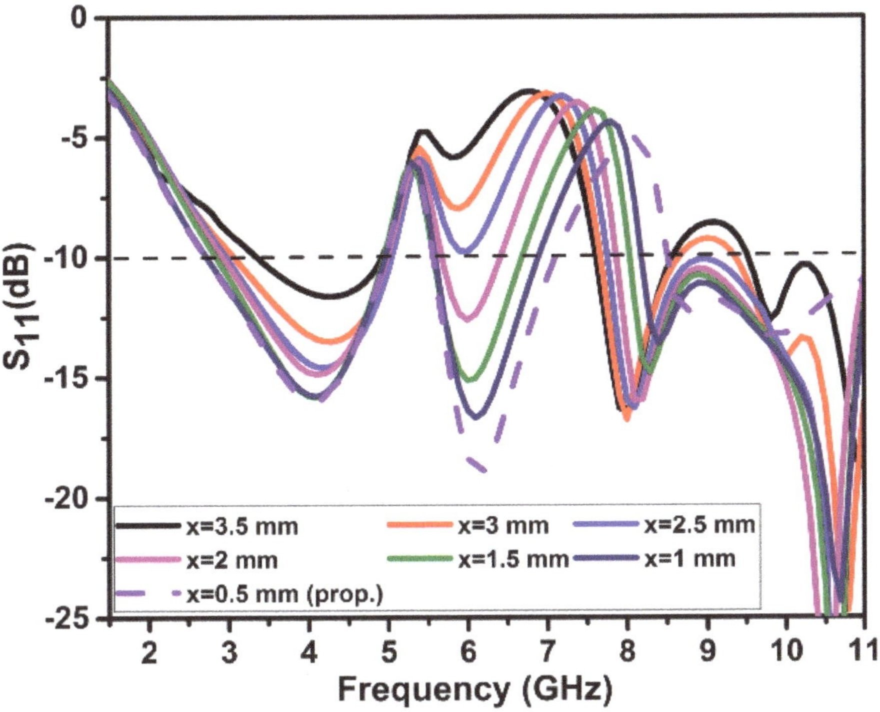

Figure 8. S_{11} for variation of width 'x'.

4. Discussion and Outcomes

The suggested antenna is subjected to a full-wave simulation using HFSS. ROHDE & SCHWARZ ZVL 13 network analyzer is used to conduct measurements.

4.1. UWB Antenna Results

The intended ultrawideband (UWB) antenna, shown in Figure 9, was fabricated according to the design.

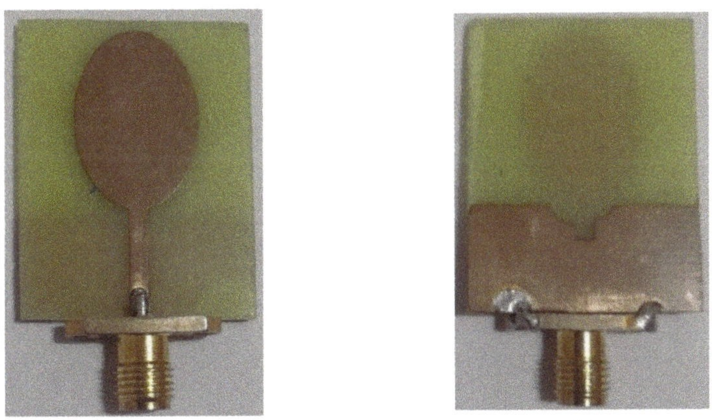

(a) (b)

Figure 9. Suggested UWB antenna: (**a**) patch; (**b**) ground plane.

4.1.1. VSWR

Figure 10 compares simulated and measured VSWR. ROHDE & SCHWARZ ZVL network analyzer measures VSWR. At VSWR < 2, the simulated impedance bandwidth is 2.8–11.2 GHz, and the operational impedance bandwidth is 2.5–11 GHz. Simulations and measurements of VSWR exhibit a minor difference. This is due to the effect of soldering, the failure to account for the SMA connector during simulation, flaws in the dielectric constant of the substrate, and manufacturing tolerances.

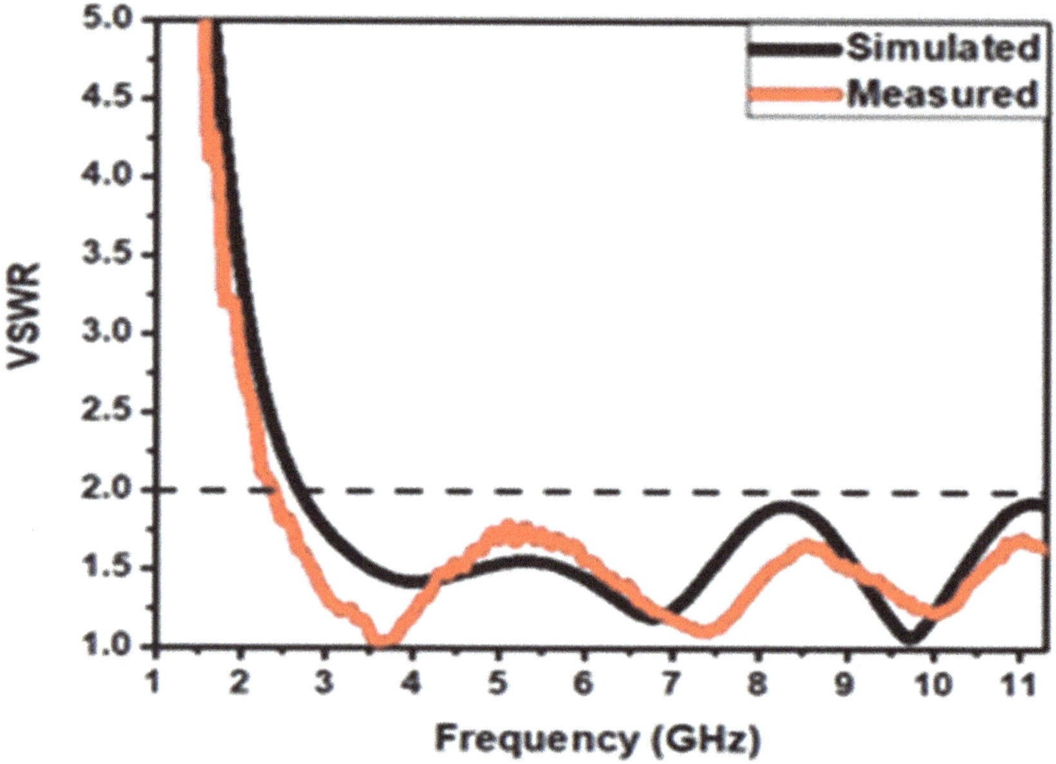

Figure 10. VSWR, simulated and measured.

4.1.2. Current Distribution

Figure 11 illustrates a simulation and study of the current distribution to illustrate the functionality of the proposed UWB antenna. At 3.6, 6.8, and 9.7 GHz, the resonance qualities are examined. Figure 11 shows how different antenna parts have varied resonance patterns for different operating modes. In Figure 11a, the lower portion of the feed is active at 3.6 GHz, where the current is highest. In Figure 11b, the feed at 6.8 GHz has the highest current. Figure 11c reveals that the lower portion of the patch and feed are most active at 9.7 GHz. Thus, the antenna resonates at 3.6, 6.8, and 9.7 GHz.

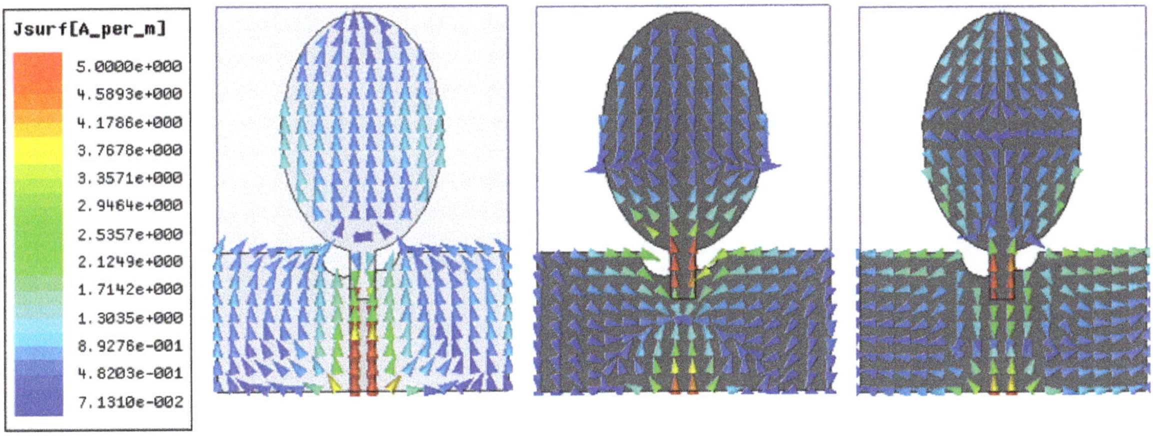

(**a**) (**b**) (**c**)

Figure 11. Current distribution at (**a**) 3.6 GHz, (**b**) 6.8 GHz, (**c**) 9.7 GHz.

4.1.3. Radiation Pattern

Figure 12 depicts the simulation of the antenna radiation pattern for resonant frequencies within the operating band. At 3.6 GHz, a bidirectional pattern is detected in the E-plane, while an omnidirectional pattern is observed in the H-plane. However, at higher frequencies (i.e., 9.7 and 12.6 GHz) the pattern is disturbed due to higher harmonics; in addition, the cross-polarization radiation magnitudes dominate the pattern. It is evident from the figure that the cross-polarization value is much lower than the co-polarization value.

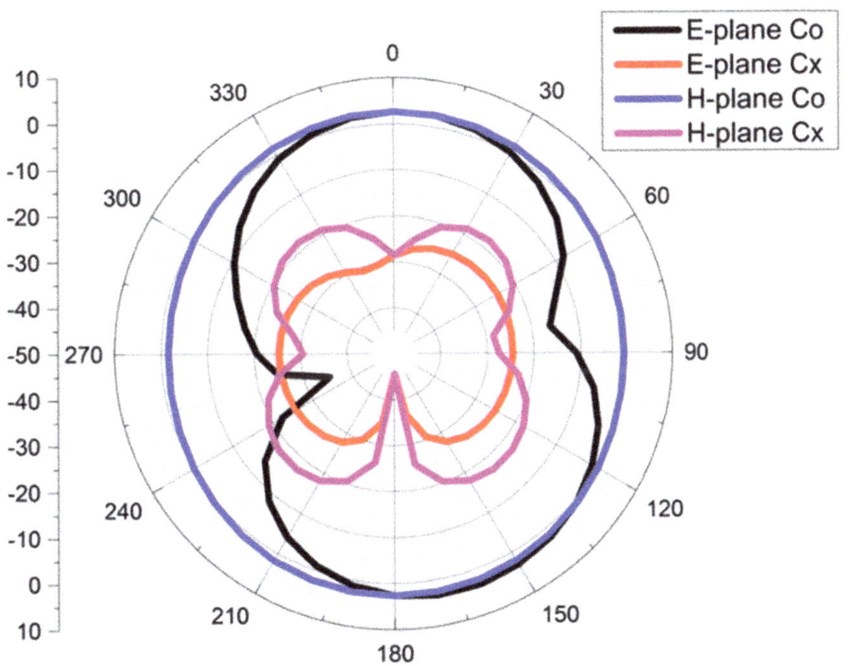

(**a**)

Figure 12. *Cont.*

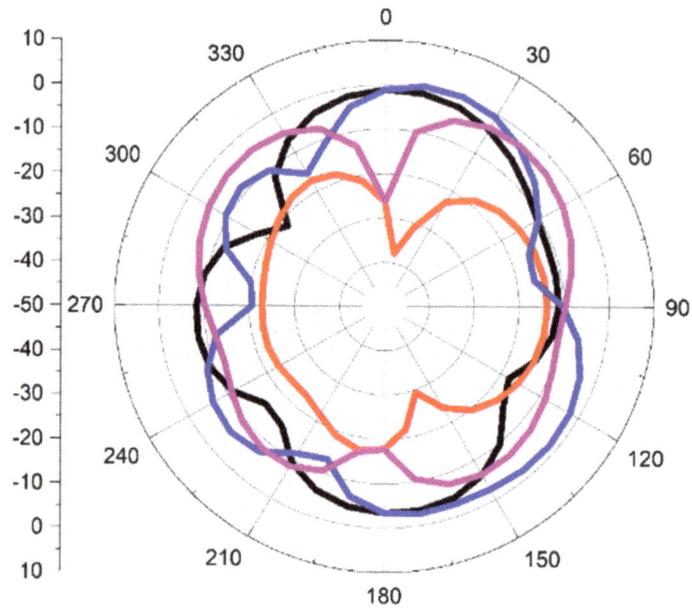

(b)

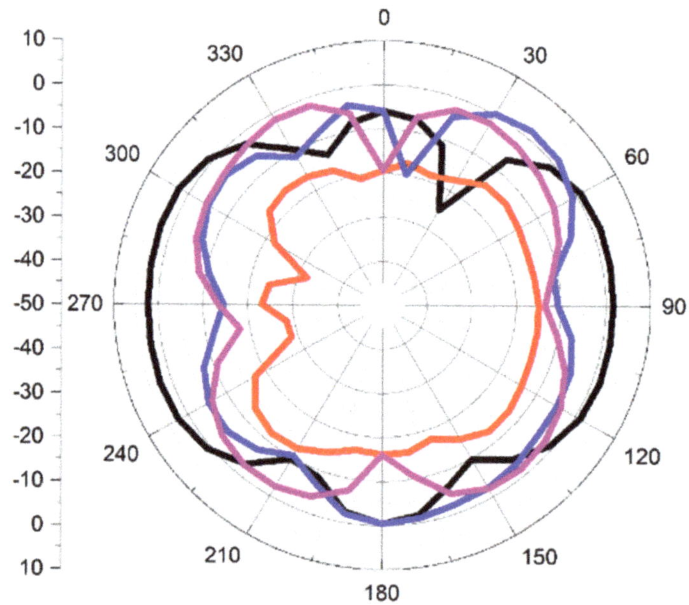

(c)

Figure 12. The simulated radiation pattern of the UWB antenna at (**a**) 3.6 GHz, (**b**) 6.8 GHz, and (**c**) 9.7 GHz.

4.1.4. Time-Domain Characteristics

The phase response, group delay, and isolation properties of the proposed UWB antenna can only be determined by a time-domain analysis. As can be seen in Figure 13, two identical antennas were placed in HFSS 100 mm apart for these tests. The time-domain study considered both forward- and side-facing scenarios.

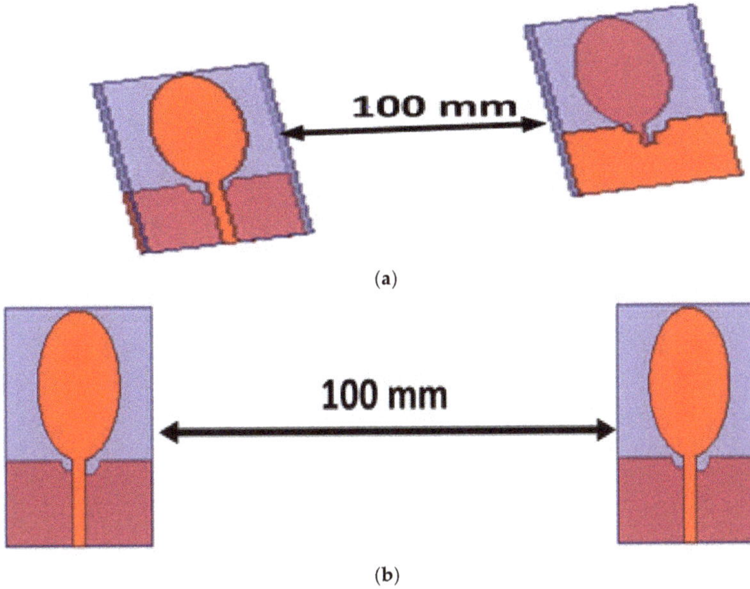

Figure 13. Time-domain analysis with orientations, viz., (**a**) frontal and (**b**) lateral.

In Equation (5), GD is the derivative of phase response to angular frequency.

$$\tau_g(\omega) = -\frac{d\varphi(\omega)}{d\omega} \quad (5)$$

The proposed UWB antenna's group delay is depicted in Figure 14. As can be seen in Figure 14, the group delay of the antenna remains relatively the same, regardless of whether it is operating in a face-to-face or side-to-side configuration.

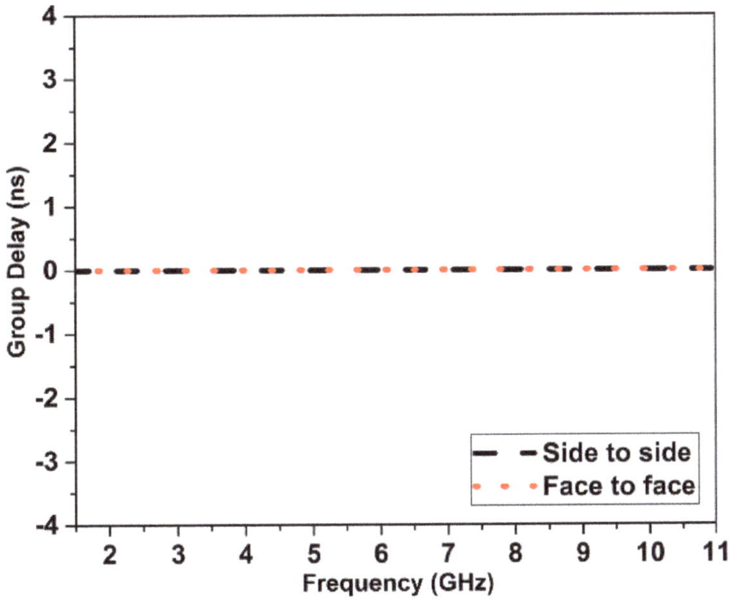

Figure 14. Group delay proposed UWB antenna.

Figure 15 shows the antenna's phase response. Linear phase variation yields desirable time-domain features. Figure 15 demonstrates that the phase fluctuation of the antenna is nearly linear during face-to-face and side-to-side operations.

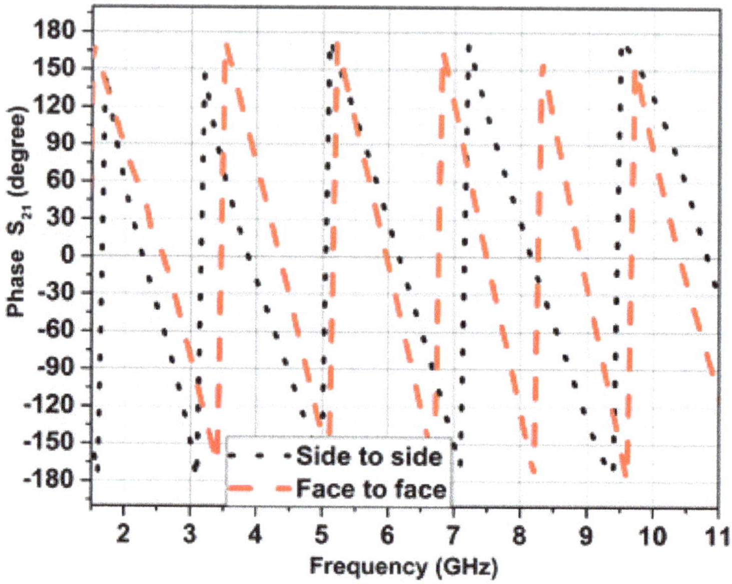

Figure 15. Phase response S_{21} of the proposed UWB design.

Isolation features of the proposed UWB antenna are depicted in Figure 16. Figure 16 shows that the antenna has good isolation characteristics ($S_{21} < -20$ dB).

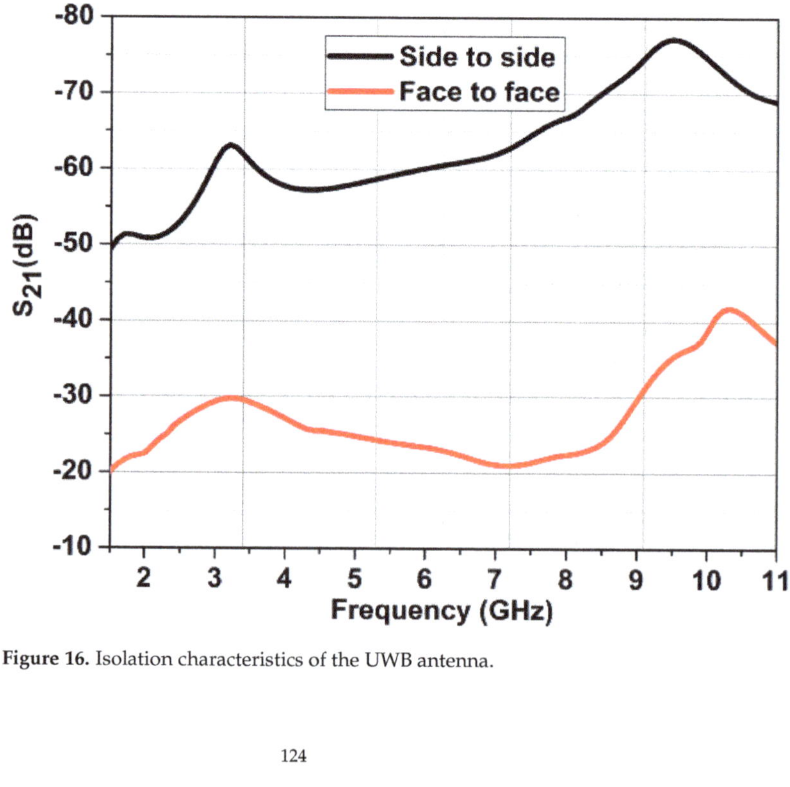

Figure 16. Isolation characteristics of the UWB antenna.

When computing the fidelity factor, a frontal orientation, as well as a lateral orientation, are both used. Both the input signals and output signals are normalized, as shown in Figure 17. The FF is computed using the normalized transmitted and received pulses in accordance with Equations (6)–(8) [22].

$$T_s^n = \frac{T_s(t)}{\sqrt{\int_{-\infty}^{\infty} |T_s(t)|^2 dt}} \tag{6}$$

$$R_s^n = \frac{R_s(t)}{\sqrt{\int_{-\infty}^{\infty} |R_s(t)|^2 dt}} \tag{7}$$

$$FF = \max \int_{-\infty}^{\infty} T_s^n(t)\ R_s^n(t+\tau)dt \tag{8}$$

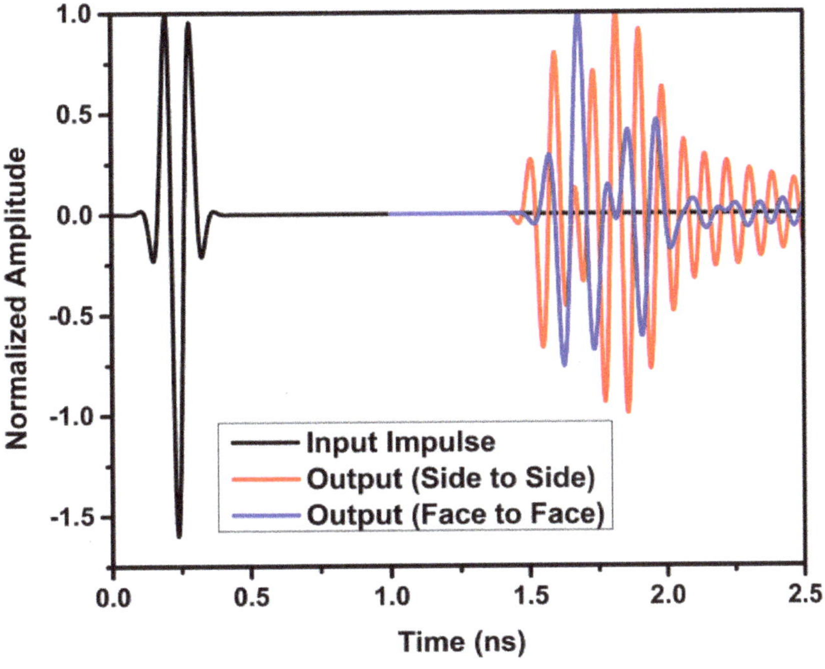

Figure 17. Fidelity factor.

Normalization is performed to ensure that only the signal's structure and not its amplitude is evaluated. The estimated fidelity factor values ensure that interference has a negligible effect on the received pulse at the receiver. The fidelity factor in frontal and lateral configurations is more than 98%. The greater value of the fidelity factor assures that the transmitted and received pulses are identical.

4.1.5. Gain

Figure 18 depicts a 3D gain plot at resonant frequencies at 3.6 GHz, 6.8 GHz, and 9.7 GHz. At 3.6 GHz, 6.8 GHz, and 9.7 GHz, the gain is 2.38 dB, 3.44 dB, and 4.85 dB.

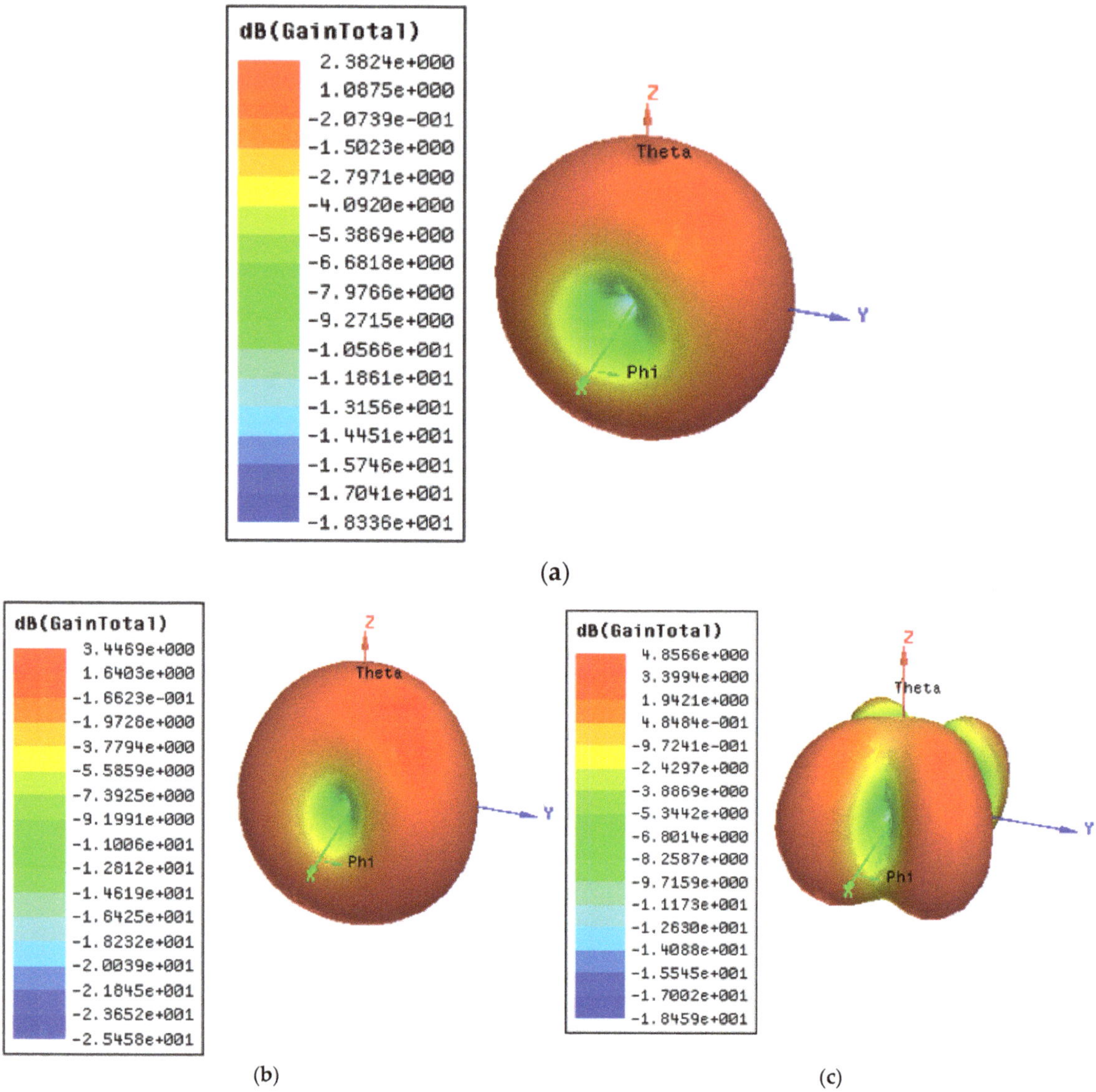

Figure 18. D gain plot at (**a**) 3.6 GHz, (**b**) 6.8 GHz, and (**c**) 9.7 GHz.

Figure 19 shows that a maximum gain of 4.9 dB is achieved for the UWB microstrip antenna.

4.2. Outcome of Ultrawideband Notch Antenna

The intended ultrawideband notch antenna, shown in Figure 20, was fabricated according to the design.

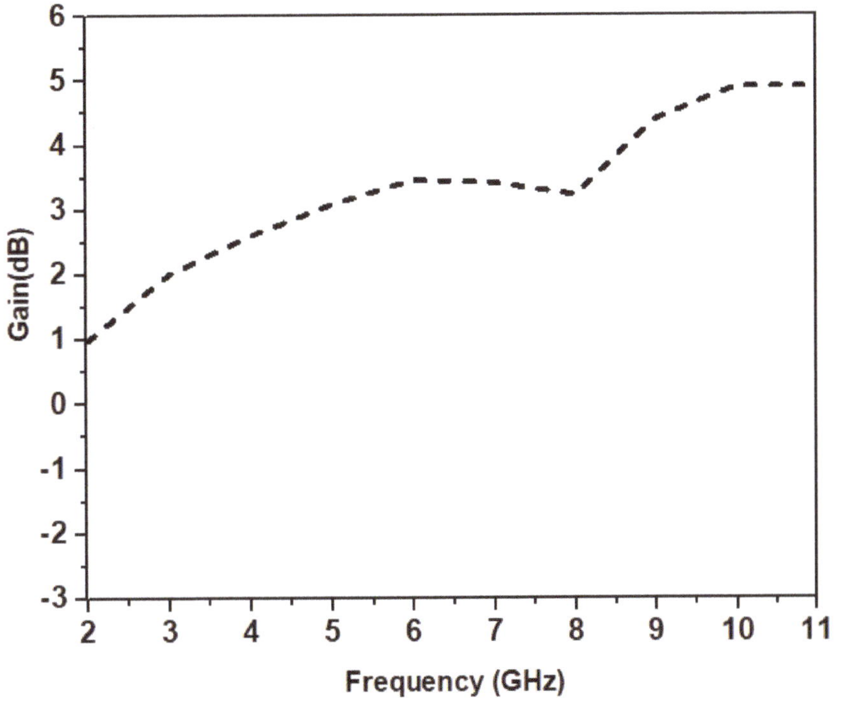

Figure 19. Gain of UWB antenna.

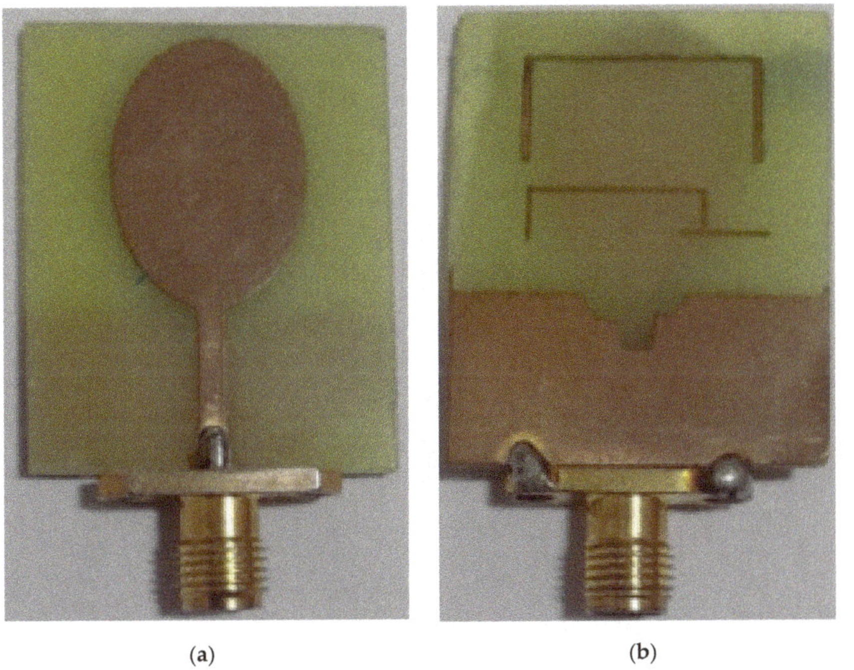

(a) (b)

Figure 20. Fabricated proposed UWB notch antenna: (**a**) patch; (**b**) ground.

4.2.1. VSWR

Figure 21 shows a comparison of the UWB notch antenna's simulated and measured VSWR. Strong rejection can be seen in the simulated VSWR plot between the WLAN band of 5–5.6 GHz and the ITU band of 7.1–8.5 GHz, with a fractional bandwidth of almost 120% (2.8–11.2 GHz). The measured VSWR plot demonstrates a high level of attenuation across the WLAN band (5.2–5.7 GHz) and the ITU band (7.2–8.5 GHz), with a fractional bandwidth of around 125%. (2.5–11 GHz). The difference in VSWR between simulated and measured values should be attributed mostly to the influence of SMA connector soldering and fabrication tolerance.

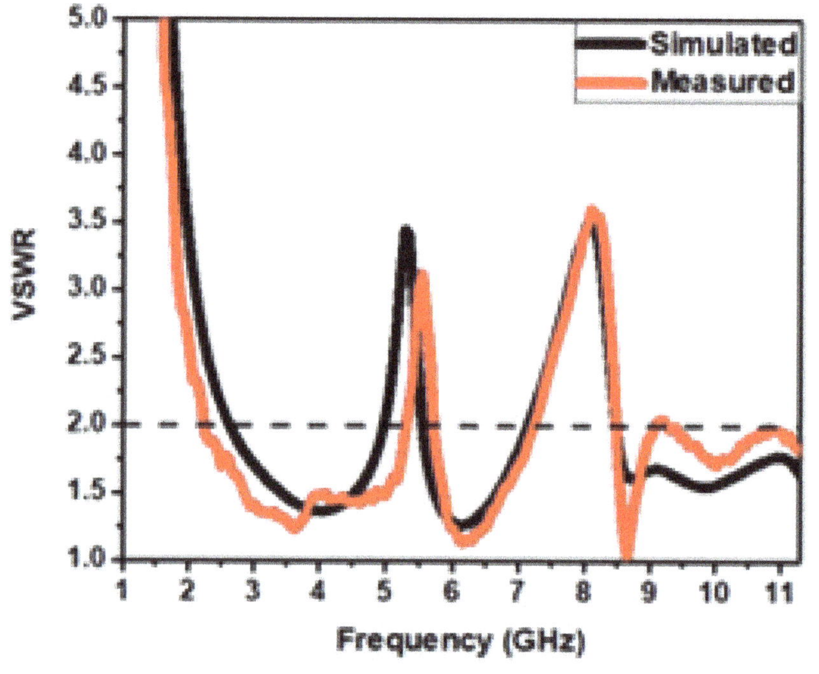

Figure 21. VSWR of UWB notch antenna, simulated and measured.

4.2.2. Current Distribution

Figure 22 depicts the surface current distributions at notched frequencies of 5.3 GHz and 8.1 GHz with parasitic resonators, illustrating the role of parasitic inverted-U-shaped and iron-shaped resonators in achieving the notched bands. At notched frequencies, surface currents on the elliptical patch flow opposite to the surface current on the ground plane, which contains parasitic elements. Accordingly, the radiation has little to no effect at the notch frequencies. It makes two notches at around 5.3 GHz and 8.1 GHz.

4.2.3. Radiation Pattern

Figure 23 depicts the simulation of the antenna radiation pattern for resonant notch frequencies within the operating band.

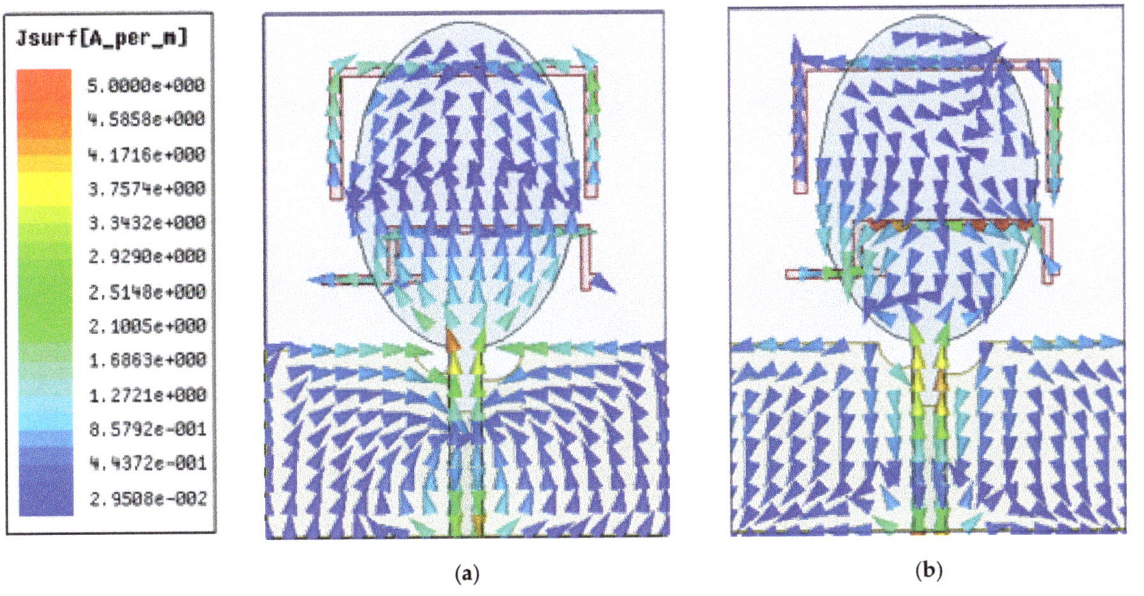

Figure 22. Current distribution at (**a**) 5.3 GHz and (**b**) 8.1 GHz.

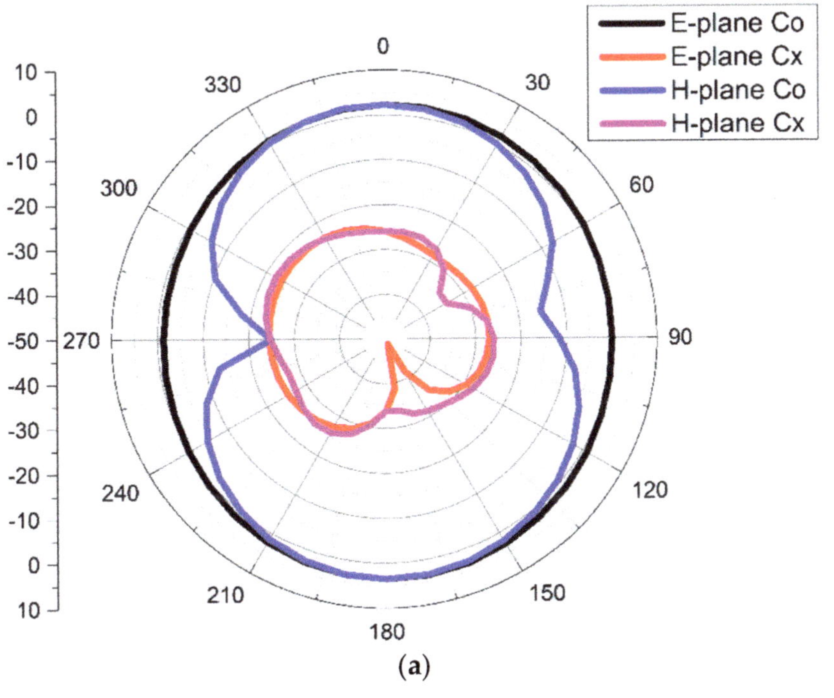

Figure 23. *Cont.*

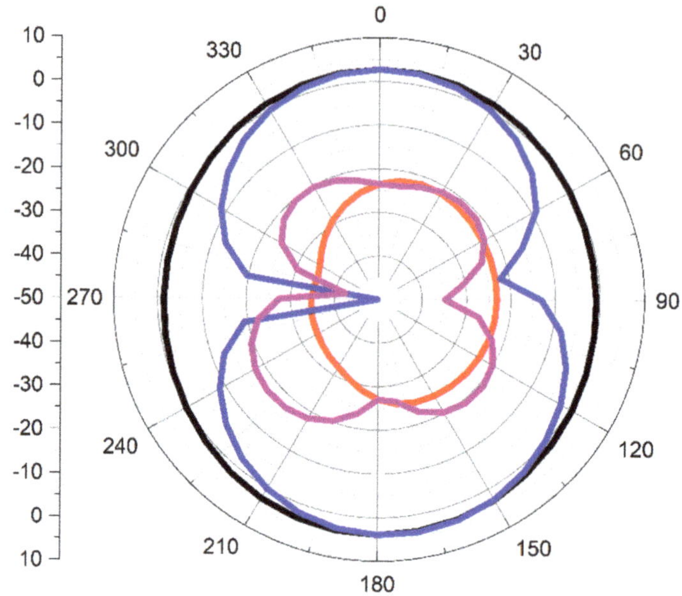

(**b**)

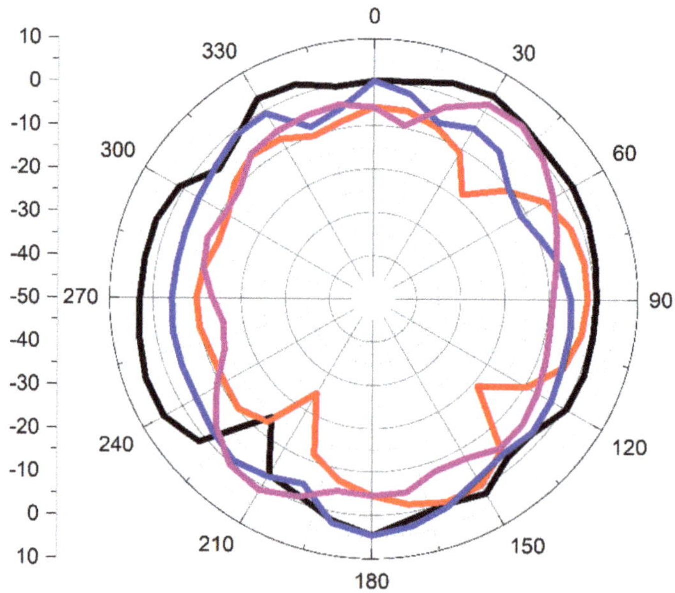

(**c**)

Figure 23. The simulated radiation pattern at (**a**) 4 GHz, (**b**) 6.2 GHz, and (**c**) 11.7 GHz.

At 4 GHz and 6.2 GHz, an omnidirectional pattern is detected in the E-plane, while a bidirectional pattern is observed in the H-plane. At 11.7 GHz, an omnidirectional pattern is seen in the E-plane and the H-plane.

4.2.4. Time-Domain Characteristics

Figure 24 illustrates analysis in the time domain, which includes group delay, isolation characteristics, and the fidelity factor for the proposed antenna. As mentioned in Section 4.1.4, these studies are carried out similarly.

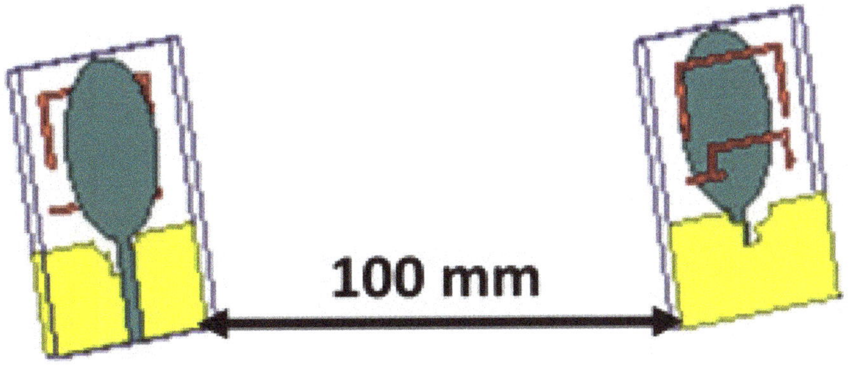

(a)

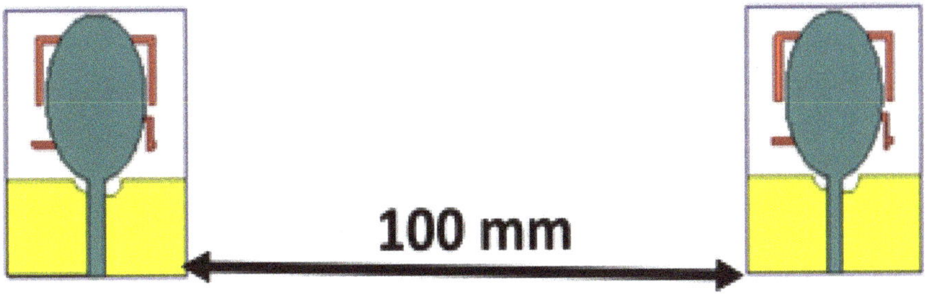

(b)

Figure 24. Time-domain analysis with orientations, viz., (**a**) frontal and (**b**) lateral.

Figure 25 shows a high delay at both notched frequencies (5.3 and 8.1 GHz), but the group delay for the face-to-face and side-to-side configurations is less than 1 ns at all other frequencies.

Figure 26 exhibits the phase variation of the input impedance of the proposed UWB notch antenna. The phase fluctuation throughout the operating band is linear, with the exception of the notched band's center frequencies, which are around 5.3 and 8.1 GHz. This difference in phase versus frequency in the notched band verifies that the signal's frequency components have a delay at the notched frequencies.

Figure 27 shows that the antenna has good isolation characteristics ($S_{21} < -20$ dB). The fidelity factor in frontal and lateral configurations is more than 99%, as shown in Figure 28. The greater value of the fidelity factor ensures that the transmitted and received pulses are identical.

For dual-band-notched UWB applications, the time domain analysis of antenna parameters shows that the suggested antenna is a good contender.

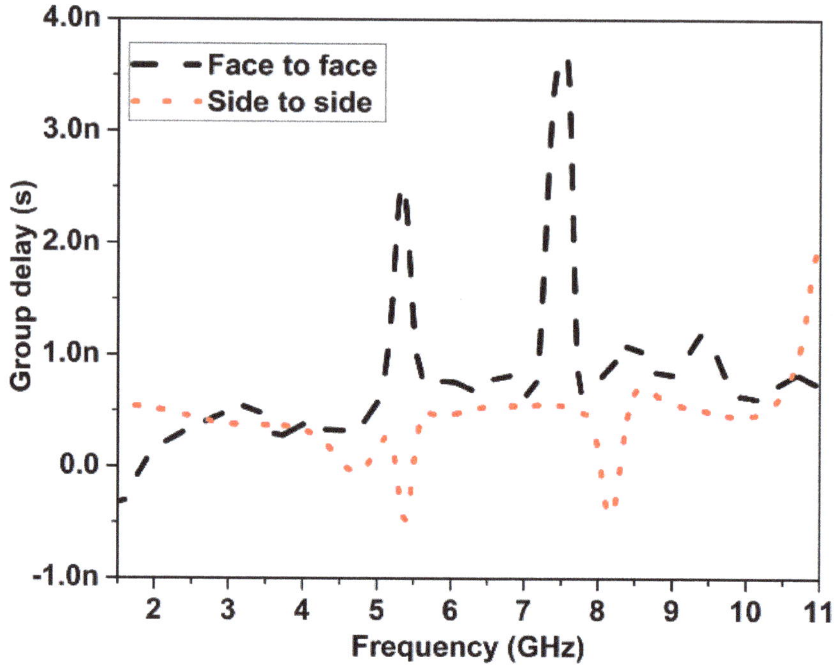

Figure 25. Group delay of proposed UWB notch antenna.

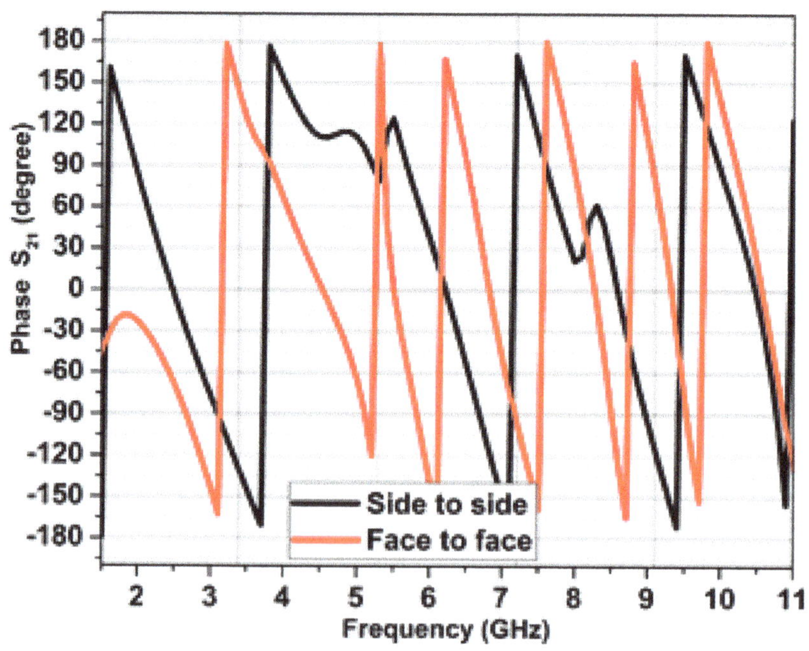

Figure 26. Phase response S_{21} of the proposed UWB notch antenna.

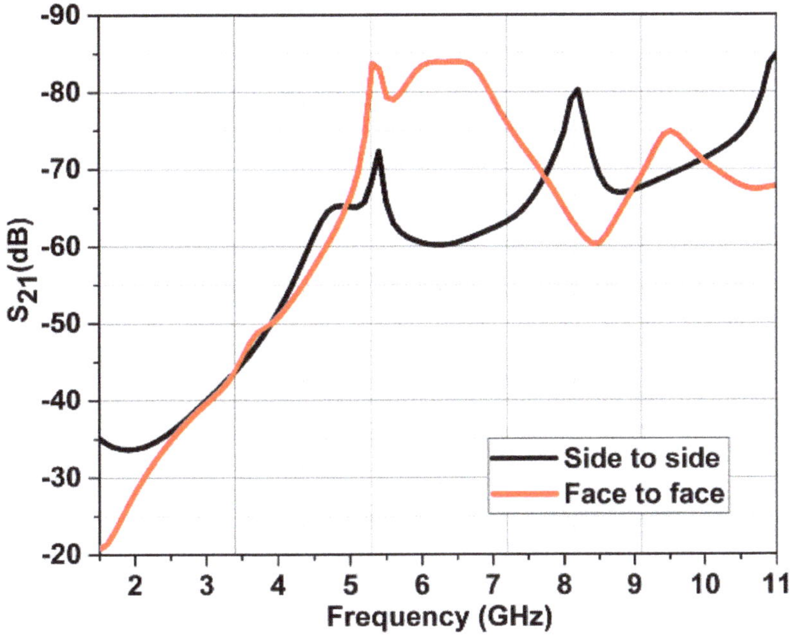

Figure 27. Isolation characteristics.

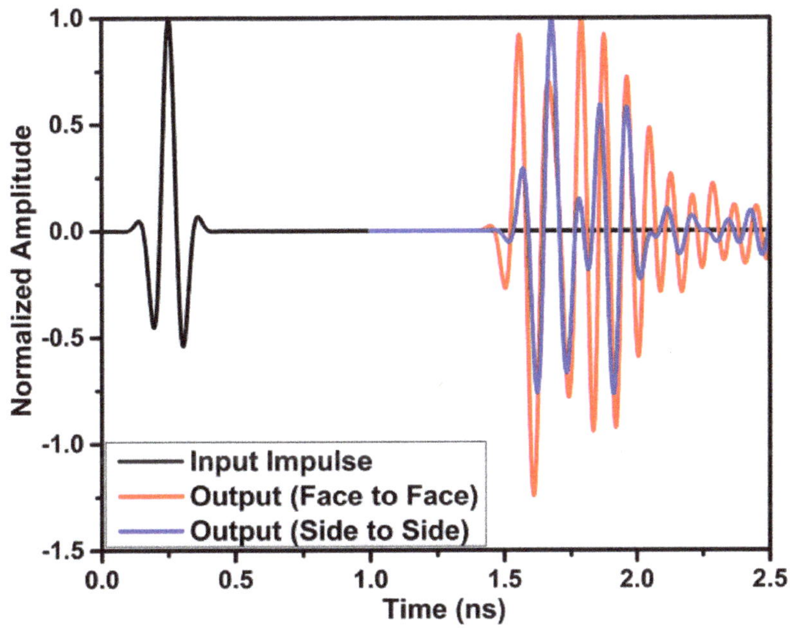

Figure 28. Fidelity factor of the UWB notch antenna.

4.2.5. Gain of UWB Notch Antenna

Figure 29 shows the total gain of the UWB notch antenna. A maximum gain of 5.9 dB is achieved for the UWB notch antenna.

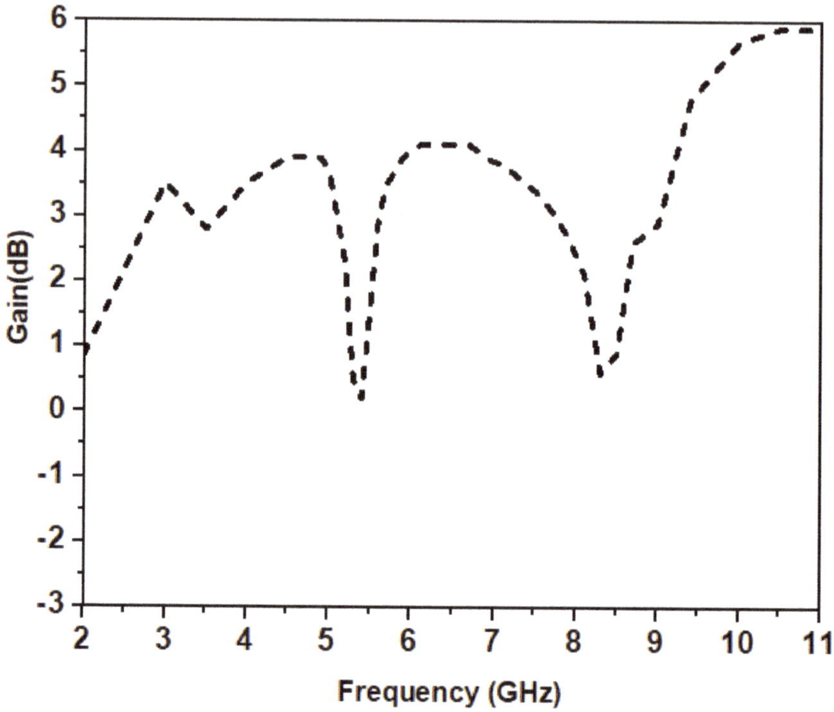

Figure 29. Gain of the UWB notch antenna.

4.2.6. Comparison Table

The expected design performance is assessed by comparing it to previously published UWB notch antenna designs, as shown in Table 2.

Table 2. Comparison of the proposed antenna with previous works.

Ref. No. & Year	Size (mm^2)	Area (mm^2)	Notch Frequency Range (GHz)	No. of Notch Frequencies	Band Notch Frequencies (GHz)		Gain (dB)
					f1	f2	
[23], 2017	$0.25\ \lambda_0 \times 0.25\ \lambda_0$	900	4.97–5.48, 5.69–5.99	2	5.2	5.8	−4.0–5.0
[24], 2018	$0.46\ \lambda_0 \times 0.55\ \lambda_0$	2100	3.3–3.6, 5–6	2	3.56	5.4	2–8.5
[25], 2019	$0.28\ \lambda_0 \times 0.26\ \lambda_0$	1720	2.3–2.5, 3.3–3.6	2	2.4	3.45	1.26–5.5
[26], 2019	$0.28\ \lambda_0 \times 0.28\ \lambda_0$	1296	2.5–2.69, 3.3–4.2	2	2.6	3.5	1.4–3.5
[27], 2020	$0.35\ \lambda_0 \times 0.46\ \lambda_0$	1200	5.5–6.2, 6.6–7.1	2	5.8	6.8	2–6
[28], 2020	$0.32\ \lambda_0 \times 0.30\ \lambda_0$	960	3.45–4.81, 5.24–6.2	2	4.4	5.8	2.7–5.1
[29], 2019	$0.30\ \lambda_0 \times 0.30\ \lambda_0$	900	5.1–6, 7.13–7.63	2	5.4	7.4	1.1–4.89
[30], 2019	$0.52\ \lambda_0 \times 0.37\ \lambda_0$	875	3.3–3.96, 4.7–5.9	2	3.6	5.3	2.4–5.5
[31], 2019	$0.45\ \lambda_0 \times 0.42\ \lambda_0$	786	5.1–5.9, 7–7.8	2	5.8	7.6	2.1–6.6
Proposed work	$0.32\ \lambda_0 \times 0.42\ \lambda_0$	768	5.2–5.7, 7.2–8.5	2	5.3	8.1	0.8–5.9

Note: λ_0 is calculated using the lowest operating frequency of the antenna.

5. Conclusions

An elliptical ultrawideband (UWB) antenna and a dual-band UWB notch antenna are described in this study. The rectangular slot on the partial ground plane is carved with two corner cuts in order to obtain UWB bandwidth. On a partial ground plane, two parasitic resonators, one of which is an inverted-U-shaped resonator, and the other of which is an iron-shaped resonator, are employed in order to obtain dual-band notch characteristics. The suggested antenna has an overall size of 24 × 32 mm^2. The suggested UWB antenna has a gain of 4.9 dB, an impedance bandwidth of 2.5–11 GHz, a linear phase response, a group delay of less than one nanoseconds, and a consistent radiation pattern. The proposed UWB notch antenna demonstrates high rejection in the WLAN band between 5.2 and 5.7 GHz, with a sharp notch at 5.3 GHz, and in the ITU band between 7.2 and 8.5 GHz, with a sharp notch at 8.1 GHz, with an impedance bandwidth ranging from 2.5 to 11 GHz. The UWB notch antenna has a linear phase, less than 1 ns group delay, and a steady radiation pattern.

Author Contributions: Conceptualization, O.P.K., T.A., P.K. (Pramod Kumar), P.K. (Pradeep Kumar) and J.A.; Methodology, O.P.K., T.A., P.K. (Pradeep Kumar) and J.A.; Software, O.P.K.; Validation, O.P.K. and T.A.; Formal analysis, O.P.K. and P.K.; Investigation, O.P.K., T.A. and P.K. (Pramod Kumar); Resources, P.K. (Pramod Kumar), P.K. (Pradeep Kumar) and J.A.; Data curation, O.P.K.; Writing—original draft, O.P.K.; Writing—review & editing, T.A., P.K. (Pramod Kumar), P.K. (Pradeep Kumar) and J.A.; Visualization, T.A. and J. Supervision, T.A., P.K. (Pramod Kumar) and J.A.; Project administration, T.A. All authors have read and agreed to the published version of the manuscript.

Funding: This research received no external funding.

Institutional Review Board Statement: Not applicable.

Informed Consent Statement: Not applicable.

Data Availability Statement: Not applicable.

Conflicts of Interest: The authors declare no conflict of interest.

References

1. Federal Communications Commission. *First Report and Order on Ultra-Wideband Technology*; FCC 02-48; Federal Communications Commission: Washington, DC, USA, 2002.
2. Kumar, O.P.; Kumar, P.; Ali, T.; Kumar, P.; Vincent, S. Ultrawideband Antennas: Growth and Evolution. *Micromachines* **2021**, *13*, 60. [CrossRef] [PubMed]
3. NXP. NXP Secure UWB Deployed in Samsung Galaxy Note 20 Ultra Bringing the First UWB-Enabled Android Device to Market. Available online: https://www.nxp.com/company/about-nxp/nxp-secure-uwb-deployed-in-samsung-galaxy-note20-ultra-bringing-the-first-uwb-enabled-android-device-to-market:NW-SECURE-UWB-SAMSUNG-GALAXY (accessed on 11 August 2020).
4. Kumar, O.P.; Kumar, P.; Ali, T. A Compact Dual-Band Notched UWB Antenna for Wireless Applications. *Micromachines* **2021**, *13*, 12. [CrossRef] [PubMed]
5. Cadence. Ultra-Wideband Antenna Applications in Communication Systems. In *Cadence System Analysis*; 2021. Available online: https://resources.system-analysis.cadence.com/blog/msa2021-ultra-wideband-antenna-applications-in-communication-systems (accessed on 11 August 2020).
6. Kumar, P.; Pai, M.M.; Ali, T. Ultrawideband antenna in wireless communication: A review and current state of the art. *Telecommun. Radio Eng.* **2020**, *79*, 929–942. [CrossRef]
7. Kumar, P.; Pathan, S.; Vincent, S.; Kumar, O.P.; Yashwanth, N.; Kumar, P.; Shetty, P.; Ali, T. A Compact Quad-Port UWB MIMO Antenna with Improved Isolation Using a Novel Mesh-Like Decoupling Structure and Unique DGS. *IEEE Trans. Circuits Syst. II Express Briefs* **2022**. [CrossRef]
8. Kumar, P.; Ali, T.; Kumar, O.P.; Vincent, S.; Kumar, P.; Nanjappa, Y.; Pathan, S. An Ultra-Compact 28 GHz Arc-Shaped Millimeter-Wave Antenna for 5G Application. *Micromachines* **2023**, *14*, 5. [CrossRef] [PubMed]
9. Kumar, O.P.; Kumar, P.; Ali, T. A Novel Arc-shaped UWB Antenna for Wireless Applications. In Proceedings of the 6th International Conference on Green Technology and Sustainable Development(GTSD), Nha Trang City, Vietnam, 29–30 July 2022; IEEE: Piscataway, NJ, USA, 2022.
10. Kumar, O.P.; Kumar, P.; Ali, T. A Novel Ultrawideband Antenna with Band Notching Facilities at WLAN, C-band, and X-band. In Proceedings of the IEEE 19th India Council International Conference (INDICON), Kochi, India, 24–26 November 2022; IEEE: Piscataway, NJ, USA, 2022.
11. Mahfuz, M.H.; Islam, M.R.; Habaebi, M.H.; Sakib, N.; Hossain, A.Z. A notched UWB microstrip patch antenna for 5G lower and FSS bands. *Microw. Opt. Technol. Lett.* **2022**, *64*, 796–802. [CrossRef]
12. Agarwal, N.; Nidadadvolu, A.; Vincent, S.; Kumar, O.P.; Ali, T. A triple band rectangular shaped patch antenna for GNSS/WiMAX applications. *AIP Conf. Proc.* **2020**, *2297*, 020012.
13. Bharadwaj, A.; Vincent, S.; Ali, T. The Floret Antenna for Terahertz Applications. *Optik* **2023**, *272*, 170335. [CrossRef]

14. Gorai, A.; Pal, M.; Ghatak, R. A Compact fractal-shaped antenna for ultrawideband and bluetooth wireless systems with WLAN rejection functionality. *IEEE Antennas Wirel. Propag. Lett.* **2017**, *16*, 2163–2166. [CrossRef]
15. Lakrit, S.; Das, S.; El Alami, A.; Barad, D.; Mohapatra, S. A compact UWB monopole patch antenna with reconfigurable Band-notched characteristics for Wi-MAX and WLAN applications. *AEU-Int. J. Electron. Commun.* **2019**, *105*, 106–115. [CrossRef]
16. Garg, R.K.; Nair, M.V.D.; Singhal, S.; Tomar, R. A miniaturized ultra-wideband antenna using "modified" rectangular patch with rejection in WiMAX and WLAN bands. *Microw. Opt. Technol. Lett.* **2021**, *63*, 1271–1277. [CrossRef]
17. Hosain, M.M.; Kumari, S.; Tiwary, A.K. Sunflower shaped fractal filtenna for WLAN and ARN application. *Microw. Opt. Technol. Lett.* **2020**, *62*, 346–354. [CrossRef]
18. Nan, J.; Zhao, J.Y.; Wang, Y. A Compact Dual Notch-Band Frequency Reconfigurable UWB Monopole Antenna. *Prog. Electromagn. Res. M* **2021**, *106*, 215–226. [CrossRef]
19. Kaur, K.; Kumar, A.; Sharma, N. Split Ring Slot Loaded Compact CPW-Fed Printed Monopole Antennas for Ultra-Wideband Applications with Band Notch Characteristics. *Prog. Electromagn. Res. C* **2021**, *110*, 39–54. [CrossRef]
20. Sohail, A.; Alimgeer, K.S.; Iftikhar, A.; Ijaz, B.; Kim, K.W.; Mohyuddin, W. Dual notch band UWB antenna with improved notch characteristics. *Microw. Opt. Technol. Lett.* **2018**, *60*, 925–930. [CrossRef]
21. Pratap Singh, A.; Khanna, R.; Singh, H. UWB antenna with dual notched band for WiMAX and WLAN applications. *Microw. Opt. Technol. Lett.* **2017**, *59*, 792–797. [CrossRef]
22. Kumar, P.; Pathan, S.; Kumar, O.P.; Vincent, S.; Nanjappa, Y.; Kumar, P.; Ali, T. Design of a Six-Port Compact UWB MIMO Antenna with a Distinctive DGS for Improved Isolation. *IEEE Access* **2022**, *10*, 112964–112974. [CrossRef]
23. Yang, B.; Qu, S. A compact integrated Bluetooth UWB dual-band notch antenna for automotive communications. *AEU-Int. J. Electron. Commun.* **2017**, *80*, 104–113. [CrossRef]
24. Jaglan, N.; Kanaujia, B.K.; Gupta, S.D.; Srivastava, S. Design of band-notched antenna with DG-CEBG. *Int. J. Electron.* **2018**, *105*, 58–72. [CrossRef]
25. Chilukuri, S.; Gogikar, S. A CPW fed denim based wearable antenna with dual band-notched characteristics for UWB applications. *Prog. Electromagn. Res. C* **2019**, *94*, 233–245. [CrossRef]
26. Puri, S.C.; Das, S.; Gopal Tiary, M. An UWB trapezoidal rings fractal monopole antenna with dual-notch characteristics. *Int. J. RF Microw. Comput.-Aided Eng.* **2019**, *29*, e21777. [CrossRef]
27. Fertas, K.; Ghanem, F.; Azrar, A.; Aksas, R. UWB antenna with sweeping dual notch based on metamaterial SRR fictive rotation. *Microw. Opt. Technol. Lett.* **2020**, *62*, 956–963. [CrossRef]
28. Puri, S.C.; Das, S.; Tiary, M.G. UWB monopole antenna with dual-band-notched characteristics. *Microw. Opt. Technol. Lett.* **2020**, *62*, 1222–1229. [CrossRef]
29. Sharma, N.; Bhatia, S.S. Design of printed monopole antenna with band notch characteristics for ultra-wideband applications. *Int. J. RF Microw. Comput.-Aided Eng.* **2019**, *29*, e21894. [CrossRef]
30. Li, J.; Sun, Y. Design of reconfigurable monopole antenna with switchable dual band-notches for UWB applications. *Prog. Electromagn. Res. C* **2019**, *96*, 97–107. [CrossRef]
31. Zhang, J.; Chen, T.; Lv, Y.; Xing, H. A practical CPW-fed UWB antenna with reconfigurable dual band-notched characteristics. *Prog. Electromagn. Res. M* **2019**, *81*, 117–126. [CrossRef]

applied sciences

MDPI

Communication

A Compact Super-Wideband High Bandwidth Dimension Ratio Octagon-Structured Monopole Antenna for Future-Generation Wireless Applications

Naineri Suguna and Senthil Revathi *

School of Electronics Engineering, Vellore Institute of Technology, Vellore 632014, India; reachsuguna@gmail.com
* Correspondence: srevathi@vit.ac.in

Abstract: A high-dimension ratio, octagonal-shaped, super-wideband (SWB) monopole antenna was proposed in this paper. The proposed antenna was composed of an octagonal-structured radiating patch with a flower-shaped slot fed by a linearly tapering microstrip line and a rectangular partial ground fabricated on a Rogers 5880 dielectric substrate, with an overall dimension of $14 \times 16 \times 0.787$ mm^3. The designed antenna exhibits SWB characteristics over the frequency range of 3.71 to 337.88 GHz at $|S11| \leq -10$ dB, VSWR < 2, a bandwidth ratio (BR) of 91.07:1, and a very high BDR of 6057.27. The proposed SWB antenna was designed, simulated, and analyzed using Ansys high-frequency structural simulator (HFSS). The simulated and measured findings have good confirmability, making them ideal for future-generation mobile networks, due to their strong radiation properties, compactness, and extremely wide bandwidth.

Keywords: polygon structure; tapered feed; super-wideband (SWB); bandwidth dimension ratio (BDR); peak gain; radiation efficiency

Citation: Suguna, N.; Revathi, S. A Compact Super-Wideband High Bandwidth Dimension Ratio Octagon-Structured Monopole Antenna for Future-Generation Wireless Applications. *Appl. Sci.* **2023**, *13*, 5057. https://doi.org/10.3390/app13085057

Academic Editor: Jaume Anguera

Received: 12 March 2023
Revised: 5 April 2023
Accepted: 5 April 2023
Published: 18 April 2023

1. Introduction

In the digital era, there is a massive demand for ultra-fast data transmission, and audio and video transmission with minimal latency in modern wireless technologies. In addition, in today's "everything and everyone is connected" world, the use of smart devices is increasing at an exponential rate, necessitating extremely wide bandwidth, compact antennas with strong radiation properties, and multiband functionality capable of supporting both short- and long-range transmission [1,2]. The Federal Communications Commission (FCC) assigned unlicensed frequency ranges of 3.1 to 10.6 GHz to ultra-wideband (UWB) technology in 2002 [3,4]. Over the past two decades, the potentials of high data rate, simplicity, low cost, low noise, high accuracy, and low power spectral density have attracted a great deal of attention. An antenna operating with a ratio bandwidth of 3.4:1 is called a UWB antenna, which enables short-range, high-speed data transmission between electronic devices. However, an antenna with a ratio bandwidth (RBW) equal to or greater than 10:1 at $|S11| \leq -10$ dB over the entire frequency range of interest is called an SWB antenna. Rumsey et al. initially developed frequency-independent SWB antennas in the late 1950s and early 1960s. These antennas consist of equiangular spiral and log-periodic structures [5]. Unlike UWB antennas, SWB antennas do not have any predefined operating frequency band standards. SWB antennas are extremely compact, have a broad bandwidth over any frequency range of the electromagnetic spectrum, and support both short- and long-range transmission. In designing SWB antennas, it is not sufficient to evaluate the antenna performance in traditional parameters such as gain and radiation patterns, etc. To verify the capability of the antenna to operate as a SWB antenna, the bandwidth dimension ratio (BDR) is introduced as an index term to determine both the compactness and wideband characteristics of an antenna. It is necessary to maintain a consistent group delay of <2 nano seconds for the entire band as well. The monopole antenna is a promising SWB antenna due to its compact dimensions, inexpensive, linear design, incredibly high BW impedance, and simplicity of integration into handheld devices [6–8].

Many wideband antennas were reported in the literature for SWB applications, such as a concentric Mickey-Mouse-shaped monopole radiator with a semi-elliptical ground for super-wideband (SWB) application, which was presented with an electrical size of 0.16 $\lambda \times$ 0.18 λ (λ lowest cutoff frequency). The antenna operates over the frequency range of 1.22–47.5 GHz and offers an FBW of 190% and a RBW of 38.9:1 with a peak gain of 0.5 to 10.3 dBi [9]. A medically designed pill-shaped microstrip patch antenna [10] with a physical size of 41 $\times$ 32 $\times$ 16 mm^3 was designed to achieve a peak gain of 8.2 dBi and operates over 2–95 GHz with BDR 3095, FBW 180%, and a low efficiency of 67%. A novel monopole antenna with a partial ground plane in the shape of a trapezoid and patches in the shape of semicircles with a tapered feed line was designed to achieve a high bandwidth dimension ratio (BDR) of 4261.007. An antenna operates in the frequency range of 1.30 to 20 GHz with an impedance bandwidth of 175.58%, a low ratio bandwidth of 15.38:1, and a low average peak gain of 4.18 dBi [11]. A compact SWB antenna was designed on a flexible Ultralam 3850 laminate substrate for 5G and IOT applications with dimensions of 60 $\times$ 40 $\times$ 0.1 mm^3 and operates in the frequency band from 1.74 to 100 GHz with a 57.47:1 ratio bandwidth (BW). A peak gain of 9.24 dB at 65.6 GHz and even lower than 0 dB of gain was observed in the designed band [12]. A circularly shaped asymmetrical dipole antenna with two dielectric substrates of low permittivity was designed with a 21.9:1 ratio bandwidth and VSWR < 2 [13]. A triple elliptical monopole antenna with overall dimensions of 27 mm $\times$ 29.5 mm was fabricated on a Roger's RT-Duroid 5880 substrate with a relative permittivity of (ε_r) = 2.2 and a loss tangent of (tan δ) = 0.0009 and covers the frequency band from 1.91 to 43.5 GHz at S11 ≤ -10 dB with a very BDR of 5761.87, a FBW > 183.17%, and a RBW of 22.77:1 [14]. A wide band scarecrow-shaped ultra-wide band (UWB) antenna using CPW and microstrip line-feeding technique was reported with a dimension of 5 $\times$ 20 $\times$ 1.6 mm^3. The IBW of a CPW-fed antenna and a microstrip-line-fed antenna was 2.51–16.48 GHz (147.13%) and 2.86–16.17 GHz (139.88%), respectively [15]. A trapezoid-shaped monopole radiator fed by a triangular, tapered feed line and a semicircular ground was presented. The electrical size was 0.16 λc $\times$ 0.27 λc and offered a SWB range of 1.42–90 GHz at S11 ≤ -10 dB, a bandwidth ratio of 63.30:1, and a VSWR $\leq$ 2 [16]. A psi-shaped circularly polarized antenna with high gain and bandwidth was investigated and found to be best suited for detection of the blind spots of a vehicle in V2V and V2R communications [17]. An elliptical-shaped fractal antenna in a tree-structured configuration for super-wideband applications was proposed. The antenna has a size of 170 $\times$ 150 mm^2 and provides an IBW of 0.65 to 35.61 GHz, a ratio bandwidth of 54.78:1, a peak realized gain of 6.51 dB, and an average gain of 3.24 dB [18].

An octagonal, ring-shaped, 40 $\times$ 40 $\times$ 1.01 mm^3 sized monopole antenna with a stub placed on the top of the right corner was developed to enhance the impedance bandwidth over a range of 2.59 to 31.14 GHz at $|S11| \leq -10$ dB [19]. A triple-notch super-wideband antenna with a bandwidth of 2.34 GHz to 20 GHz was designed and fabricated for Bluetooth and LTE 2600 bands, and achieved a gain of 4.98 dBi and radiation efficiency of 89% [20]. An offset elliptical ring patch antenna with a tapered feed line was developed. The proposed antenna has a ratio bandwidth of 34.63:1, a bandwidth dimension ratio of 1732, and a peak gain of 5.81 dBi over the functional frequency range of 2.31–40.0 GHz [21]. A compact bulb-model planar super-wideband SWB antenna was designed with the dimensions of 35 $\times$ 30 $\times$ 0.8 mm^3 and works in the frequency range of 2.8–40 GHz with a RBW of 14.28:1. The antenna has a FBW of 173.8% and a BDR of 1904. A FSS was placed on top of the radiating patch with an air gap of 19 mm to enhance gain over the specified SWB range [22]. A Sierpinski fractal antenna was designed with a patch in the shape of a hexagon and coplanar waveguide feeding. The bandwidth ratio of 11:1 was achieved over an impedance bandwidth of 3.4–37.4 GHz [23]. A star–triangular fractal monopole antenna feed with microstrip-fed and semielliptical ground plane was investigated with a footprint of 20 mm × 20 mm × 1 mm^3 and exhibited a super-wideband range of 1–30 GHz with VSWR < 2 [24]. The SWB antennas reported in [9–24] have limited FBW, BDR, and radiation performance. Therefore, the major issues in the design of SWB antennas are miniaturization, wide impedance bandwidth, and a high BDR.

A compact super-wideband antenna was presented and analyzed in this article. It consists of an octagonal radiating patch fed by a linearly tapered microstrip line and the partial ground plane. The proposed antenna achieved a super-wideband frequency range of 3.71–337.88 GHz, a fractional bandwidth of 195.65%, and a very high bandwidth dimension ratio of 6057.27, with an overall size of 14 mm × 16 mm. VSWR, efficiency, and radiation patterns were also investigated to support SWB applications.

2. Antenna Design Methodology

The geometrical parameters of the patch antenna are computed using conventional mathematical formulas [25,26], and then applied to area of an octagon to derive its dimensions. An octagonal patch antenna is constructed on the basis of a circular patch antenna. The area of the octagonal and circular geometry is nearly equal, as shown in Figure 1. The dimensions of the circular and octagonal shapes are determined using Equations (1)–(4) and optimized for the better antenna performance.

$$\pi {a_e}^2 = 2\left(1+\sqrt{2}\right)s^2 \tag{1}$$

where 'a_e' is effective radius of circular patch due to fringing effect

$$a_e = a\left[1+\frac{2h}{\pi a \varepsilon_r}\left\{\ln\left(\frac{\pi a}{2h}\right)+1.7726\right\}\right]^{1/2} \tag{2}$$

'a' is the actual radius of the circular patch and is derived as

$$a = \frac{F}{1+\frac{2h}{\pi F \varepsilon_r}\left[\ln\left(\frac{\pi F}{2h}\right)+1.7726\right]^{1/2}} \tag{3}$$

where

$$F = \frac{8.791 * 10^9}{f_r\sqrt{\varepsilon_r}} \tag{4}$$

The proposed monopole antenna with various evaluations are shown in Figure 2. The electrical dimension of the antenna is 0.16 λc × 0.18 λc, where the wavelength λc is the lowest operating frequency or physical size of 14 mm × 16 mm. It is printed on Rogers RT/Duroid 5880TM dielectric substrate with a permittivity (εr) of 2.2 and a height (h) of 0.787 mm. The physical dimensions of substrate are Ws × Ls, where Ws = 14 mm and Ls = 16 mm. The presented antenna has an octagonal-shaped radiating patch, tapered feed microstrip line, and partial ground plane. To achieve optimal matching, the width of the feed line is reduced linearly from W2 = 2 mm to W1 = 1.5 mm, while the length of the feed line Lf = 6.3 mm remains constant. The ground plane length Lg = 6 mm, and the radius of the circle patch (R) = 4.5 mm, from which the octagonal sides are derived. The edge of the octagon, S = 3.45 mm, and the elliptical slots D1, D2, D3, and D4 are used in the patch to construct a flower-shaped slot in the octagonal radiating patch. A very wide frequency range is also dependent on the geometry of the ground, and the time of the current path determines the lowest cut off frequency. As a result, rectangular ground becomes a partially ground plane.

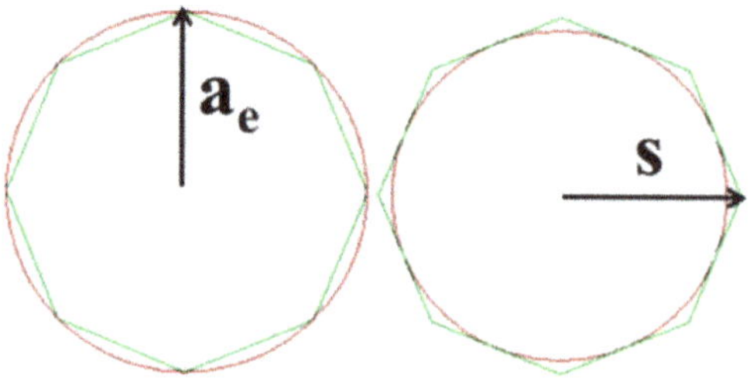

Figure 1. Area correlation between circular and octagon geometry.

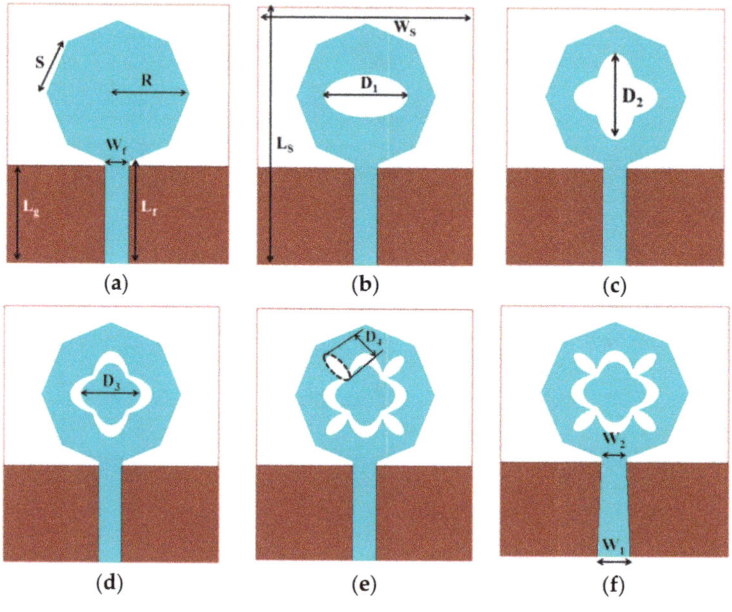

Figure 2. Geometry of SWB antenna with different design evaluations (**a**) iteration 1, (**b**) iteration 2, (**c**) iteration 3, (**d**) iteration 4, (**e**) iteration 5, (**f**) iteration 6.

Evaluations of SWB Monopole Antenna Configuration

Different evaluations of the octagonal structured proposed monopole antenna with a partial ground plane are shown in Figure 2. The octagonal radiating patch designed on Rogers RT/Duroid 5880™ material with partial ground geometry is shown in Figure 2a. An elliptical aperture is cut into the octagon patch to enhance its impedance bandwidth, as shown in Figure 2b. The Babinet principle is applied during this process. The elliptical slot is converted into a plus-shaped elliptical slot and embedded into the patch, as shown in Figure 2c. When the self-similar modified plus structure is merged with the preceding geometry, a new ring-slotted octagon patch is formed, as shown in Figure 2d. The octagonal structure that is carved into the flower-shaped slot is shown in Figure 2e. Furthermore, IBW and BDR are improved by switching from a rectangular 50 Ω transmission feed line to a tapered feed line, as shown in Figure 2f. The dimensions of the evaluated geometries are listed in Table 1.

Table 1. Novel self-similar slotted monopole antenna with partial ground plane dimensions given in mm.

Design Parameters	Dimensions	Design Parameters	Dimensions
L_f	6.3	D_1	5.45
W_f	1.5	D_2	5.45
L_g	6	D_3	3.8
S	3.45	D_4	1.8
R	4.5	W_1	1.5
L_s	16	W_2	2
W_s	14		

3. Results and Discussion

3.1. Electromagnetic Characteristics

The proposed super-wideband monopole antenna is analyzed in terms of reflection coefficient (S11), voltage standing-wave ratio (VSWR), and input impedance (Z_{in}), as shown in Figures 3–5.

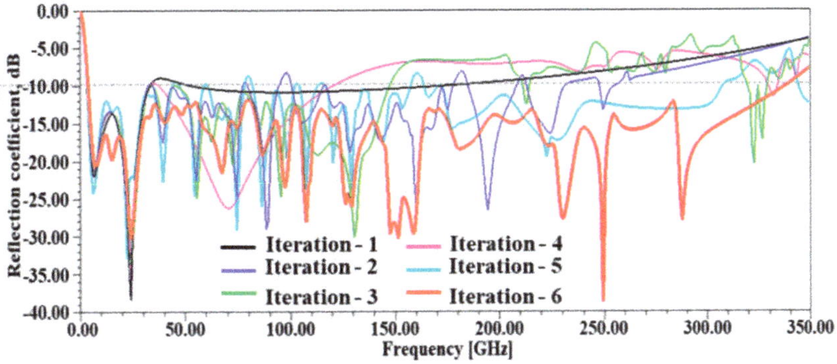

Figure 3. Simulated reflection − coefficient of proposed antenna.

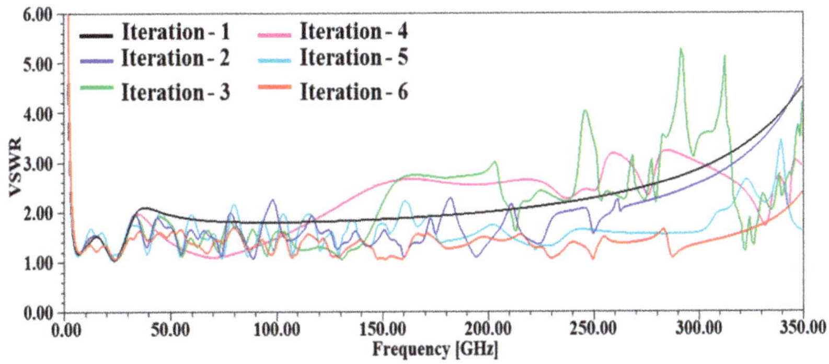

Figure 4. VSWR of the proposed antenna.

Figure 3 demonstrates the reflection coefficient of the proposed antenna for all the iterations from 1 to 6, whose electrical characteristics are summarized in Table 2. In iteration 1, an eight-sided polygon radiating element of the copper conductor layer with a partial ground plane is developed, and this antenna has a maximum reflection coefficient (S11) of −38.41 dB at resonance frequency of 24.1 GHz and an impedance bandwidth (IBW) of 29.69 GHz over a frequency range of 3.49–33.18 GHz with S11 ≤ -10 dB, as shown in Figure 3. In iteration 2, an elliptical slot is etched into the polygon structure to obtain wider bandwidth characteristics, while the ground plane remains unchanged from the first step. As shown in Figure 3, this antenna covers multiple bands in the desired spectrum, and the corresponding IBWs are 28.67 GHz, 42.73 GHz, 14.82 GHz, 15.56 GHz, 61.1 GHz, 23.58 GHz, and 18.52 GHz, as listed in Table 2. The antenna design has been modified to avoid notches by incorporating an elliptical plus-shaped slot into the octagonal patch design. Iteration 3, provides bandwidths of 3.54–33.69 GHz and 45.28–147.10 GHz, with respective FBWs of 161.96% and 105.85%. There is a significant enhancement in both the wide and fractional bandwidths, though the notches are not completely eliminated, as shown in Figure 3. In iteration 4, a plus-shaped ring is etched into a conventional octagon, which enhances the fractional bandwidth and bandwidth characteristics illustrated in Figure 3. These characteristics restrict the fractional bandwidth of two broad bands to 161.36% and 105.76%, respectively. In iteration 5, the radiating element is optimized by integrating the minor elliptical slots and the ring slots to resemble a flower structure. The bandwidth is increased to cover the frequency range of 3.37 GHz to 307.24 GHz with minor notch bands at 43.80–47.15 GHz, 58.82–60.47 GHz, 77.98–81.87 GHz, 101.92–103.88 GHz, 113.03–116.49 GHz, and 157.39–171.96 GHz with the modified structure. In addition to the notch bands, the iterative structure exhibits a huge bandwidth, which covers a wide range of wireless applications. The main objective of the proposed structure is to cover a wide band spectrum suitable for a diverse range of wireless applications such as industrial, scientific, and medical (ISM), radar, satellite, navigation, sensing, explosive detection, short-range indoor applications, etc. Iteration 6 transforms a 50 Ω transmission line into a tapered-fed structure with a partial ground plane. The proposed monopole structure with optimized geometry, as shown in Figure 3, has an operating range of 3.71–337.88 GHz at $S11 \leq -10$ dB and an impedance bandwidth of 334.17 GHz, a bandwidth ratio (BR) of 91.07, and a fractional bandwidth (FBW) of 195.65%, as stated in Table 2. The proposed antenna achieves a very high bandwidth dimension ratio (BDR) of) of 6057.27 due to the novel structure. The simulated VSWR graph of design evaluations from conventional polygons to proposed geometry is shown in Figure 4. A low VSWR across the operating band indicates low mismatch losses, low reflected power, and effective impedance matching. The mathematical relation between mismatch loss (ML) and VSWR is shown in Equation (5).

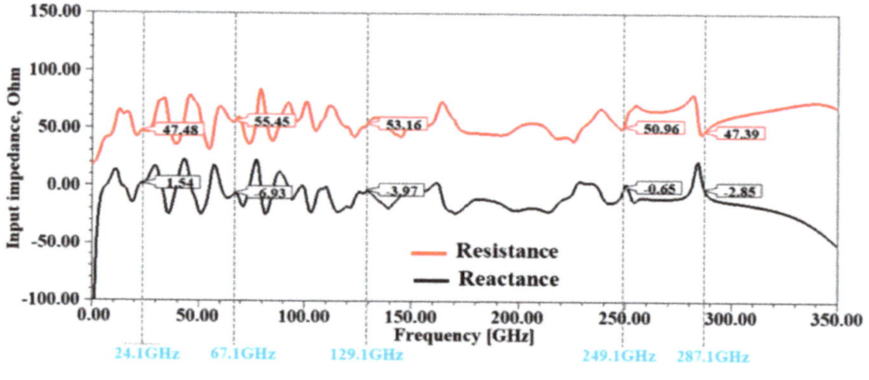

Figure 5. Simulated input impedance of designed antenna: real and imaginary part.

$$ML = -10\log\left[1 - \left[\frac{VSWR - 1}{VSWR + 1}\right]^2\right] \quad (5)$$

Table 2. Electrical characteristics summary of designed antenna in various stages.

S. No	Design	Operating Bandwidth, GHz	Impedance Bandwidth, GHz	Bandwidth Ratio, BR	Fractional Bandwidth, FBW %	VSWR
1	Iteration 1	3.49–33.18	29.69	9.507	161.93%	1.03
		3.43–32.10	28.67	9.35	161.38%	1.06
		34.32–77.05	42.73	2.24	76.73%	1.14
		79.37–94.19	14.82	1.18	17.07%	1.08
2	Iteration 2	100.27–115.83	15.56	1.15	14.40%	1.44
		116.65–177.75	61.1	1.52	41.50%	1.09
		184.63–208.21	23.58	1.12	12.00%	1.09
		212.68–231.20	18.52	1.08	8.34%	1.35
3	Iteration 3	3.54–33.69	30.15	9.51	161.96%	1.04
		45.28–147.10	101.82	3.24	105.85%	1.06
4	Iteration 4	3.54–33.11	29.57	9.35	161.36%	1.02
		37.17–120.58	83.41	3.24	105.75%	1.10
5	Iteration 5	3.37–43.80	40.43	12.99	171.42%	1.04
		47.15–58.82	11.67	1.24	22.08%	1.11
		60.47–77.98	17.51	1.29	25.29%	1.07
		81.87–101.92	20.05	1.24	21.81%	1.23
		103.88–113.03	9.15	1.08	8.43%	1.13
		116.49–157.39	40.9	1.35	29.86%	1.13
		171.96–307.24	135.28	1.78	56.46%	1.23
6	**Iteration 6**	**3.71–337.88**	**334.17**	**91.07**	**195.65%**	**1.02**

The significant enhancements in operating bandwidth, impedance bandwidth, bandwidth ratio, fractional bandwidth, and VSWR for each antenna iteration are shown in Table 2. The values are computed using Equations (6), (7), (8), and (9), respectively.

Impedance bandwidth (IBW):

$$IBW = f_h - f_l \tag{6}$$

Bandwidth ratio (BR):

$$BR = f_h / f_l \tag{7}$$

Fractional bandwidth (FBW):

$$FBW = \frac{2 * (f_h - f_l)}{(f_h + f_l)} * 100 \tag{8}$$

where f_l is the low cut-off frequency, and f_h is the high cut-off frequency, and VSWR of iterations 1–6.

BDR is an essential parameter in SWB antennas because it implies the antenna compactness and wider bandwidth. BDR denotes the proportion of bandwidth that an antenna can provide per unit electrical area. BDR is defined mathematically [5] as Equation (9).

$$BDR = \frac{Bandwidth\ \%}{\lambda_{length}\ X\ \lambda_{width}} \tag{9}$$

where λ is the lowest cutoff frequency of operating spectrum.

The input impedance (Z_{in}) characteristics of the proposed tapered-fed monopole element with a partial ground plane are shown in Figure 5. At resonant frequencies of 24.1 GHz, 67.1 GHz, 129.1 GHz, 249.1 GHz, and 28 GHz, the resistance (R_{in}) and reactance (X_{in}) values are 47.48 + j1.54 Ω, 55.45 − j6.93 Ω, 53.16 − j3.97 Ω, 50.96 − j0.65 Ω, and 47.39 − j2.85 Ω, respectively, and are listed in Table 3.

Table 3. Input impedance of proposed antenna at selected resonant frequencies.

S. No	Resonant Frequency	Resistance, R_{in}	Reactance, X_{in}	Input Impedance, Z_{in}
1	24.1 GHz	47.48 Ω	1.54 Ω	47.48 + j1.54 Ω
2	67.1 GHz	55.45 Ω	−6.93 Ω	55.45 − j6.93 Ω
3	129.1 GHz	53.16 Ω	−3.97 Ω	53.16 − j3.97 Ω
4	249.1 GHz	50.96 Ω	−0.65 Ω	50.96 − j0.65 Ω
5	287.1 GHz	47.39 Ω	−2.85 Ω	47.39 − j2.85 Ω

3.2. Radiation Characteristics

The proposed antenna is simulated using the Ansys HFSS simulator tool, which produces peak gains of 2.31 dBi at 24.1 GHz, 6.11 dBi at 67.1 GHz, 12.00 dBi at 129.1 GHz, 14.14 dBi at 249.1 GHz, and 14.29 dBi at 287.1 GHz, as shown in 3D gain polar plots in Figure 6. Also, 2D far-field patterns in E-plane and H-plane are shown in Figure 7. At low frequencies, the current is distributed uniformly across the ground plane and patch (Figure 6a,b). As a result, radiation patterns in the E and H planes are omnidirectional. At mid- and higher frequencies, higher-order modes are generated, and the current density on the radiator is not as evenly distributed. Therefore, the radiated modes are more prone to high radiation interference or minor lobes. As illustrated in Figure 6c–e, deterioration in the omnidirectional nature of the radiation pattern begins with an increase in frequency. Various techniques, such as periodic structures, defected ground structures (DGS), parasitic elements, and split ring resonators, can be employed to suppress these undesired radiated modes, thereby improving the performance of the proposed antenna over the SWB range.

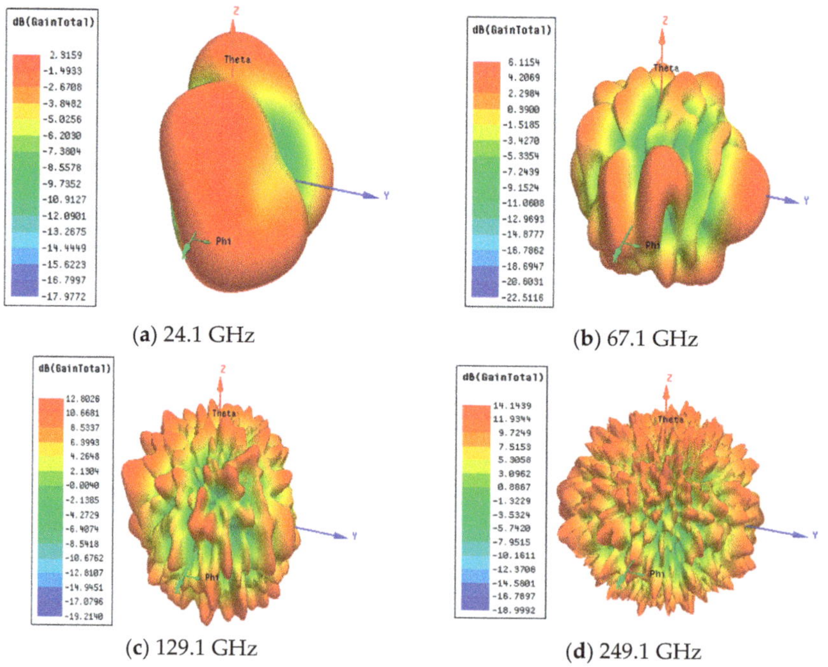

(**a**) 24.1 GHz (**b**) 67.1 GHz

(**c**) 129.1 GHz (**d**) 249.1 GHz

Figure 6. *Cont.*

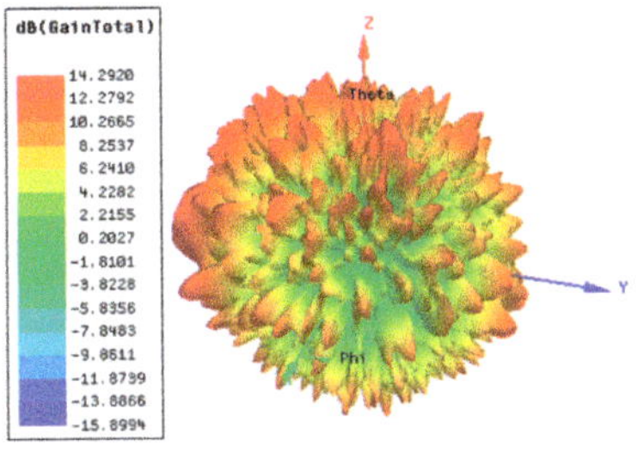

(**e**) 287.1 GHz

Figure 6. The 3D gain polar plots of the final iterative design of the proposed tapered-fed structure with partial ground plane.

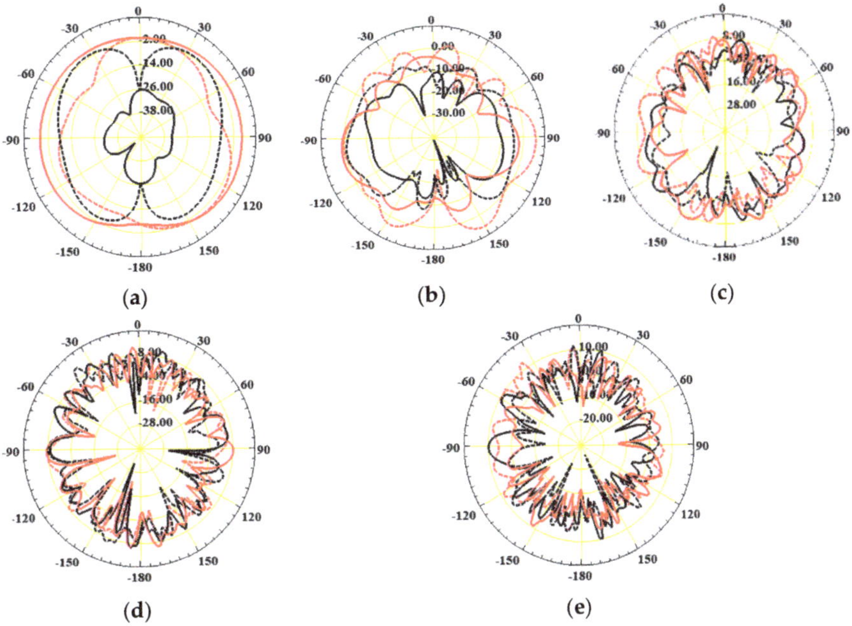

Figure 7. The 2D far-field patterns in E-plane and H-plane: (**a**) 24.1 GHz, (**b**) 67.1 GHz, (**c**) 129.1 GHz, (**d**) 249.1 GHz, (**e**) 287.1 GHz (red solid—co-polarized E-plane; black solid—cross-polarized E-plane; red dotted line—co-polarized H-plane; black dotted line—cross-polarized H-plane).

3.3. Surface Current Analysis

The performance of a super-wideband (SWB) antenna is analyzed using surface current dissipation, as shown in Figure 8a–e. Current is more concentrated around the feedline region and the edges of the minor elliptical slots, at the resonance frequencies of 24.1 GHz, 67.1 GHz, 129.1 GHz, 249.1 GHz, and 287.1 GHz of the final iteration of the proposed antenna, as shown in Figure 8a. Current flows through the feedline and concentrates at the slot edges depending on the phase angle, as shown in Figure 8b. The magnitude of the current at the feedline is higher at 129.1 GHz, as shown in Figure 8c. At higher resonant frequencies of 249.1 GHz and 287.1 GHz, more current flows to the feed line and ground plane, as shown in Figure 8d,e. The iterations in the antenna design produce a fringing field, which improves the coupling mechanism between the radiating layers. As a result, the impedance of the proposed monopole antenna increases.

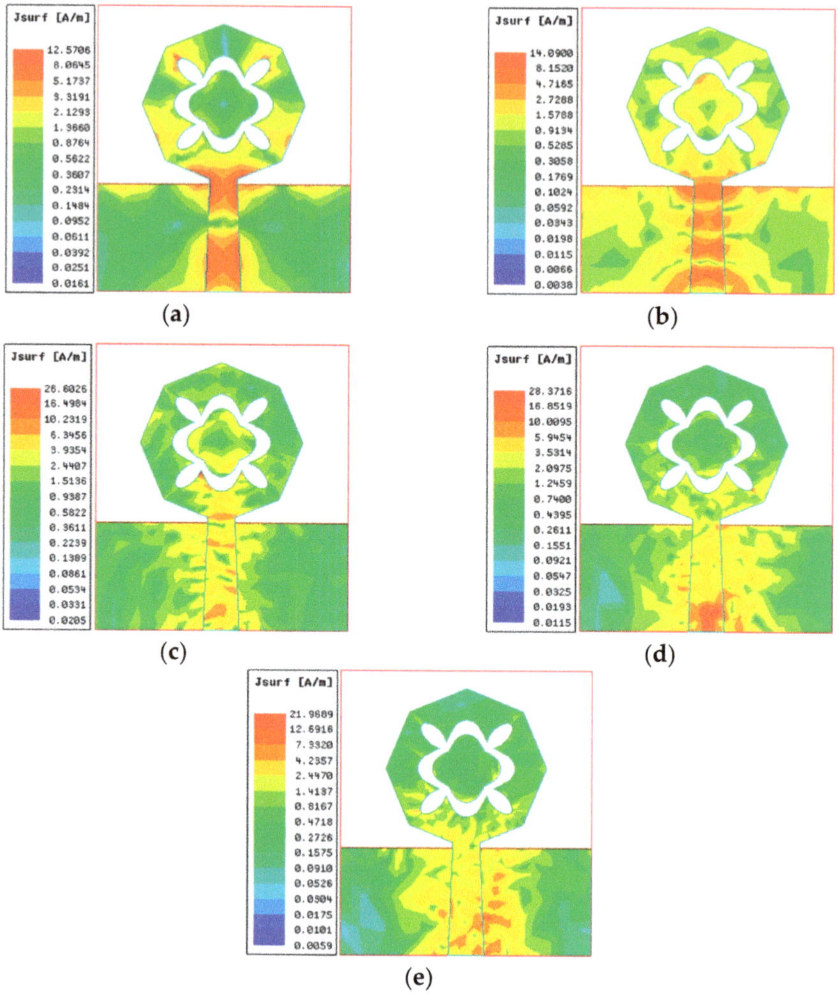

Figure 8. The surface current distribution of the designed antenna for (**a**) 24.1 GHz, (**b**) 67.1 GHz, (**c**) 129.1 GHz, (**d**) 249.1 GHz, (**e**) 287.1 GHz.

3.4. Gain and Radiation Efficiency Characteristics

The simulated gain (G) and radiation efficiency (η) of a SWB monopole antenna are shown in Figures 9 and 10. The proposed design has a peak-realized gain of 18.01 dB and a radiation efficiency of 62–95% in the super-wideband frequency range of 3.71–337.88 GHz. As shown in Figure 9, there is a minor discrepancy between realized gain and total gain up to 255.56 GHz, and an exponential drop at higher operating frequencies due to mismatch losses and reflection losses at the input port. Similarly, as shown in Figure 10, these losses affect the radiation efficiency in the same manner. This high gain is achieved by etching a flower-shaped slot into a radiator and employing a partial ground and tapered feed technique.

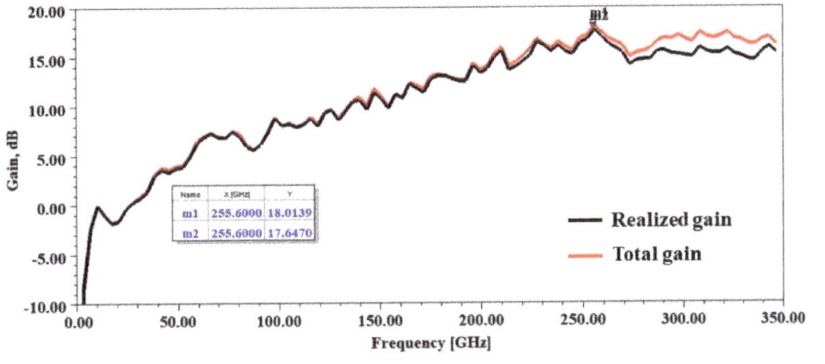

Figure 9. Gain–characteristics of the suggested antenna.

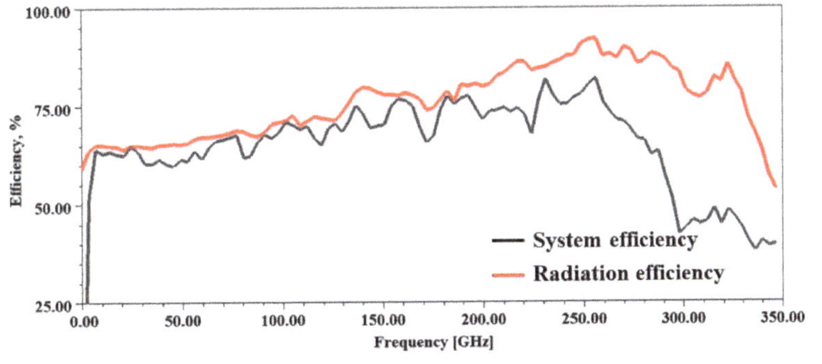

Figure 10. Efficiency characteristics of the suggested antenna.

3.5. Fabricated Prototype and Measured Results

The front and rear views of the fabricated prototypes of a proposed monopole SWB antenna printed using a DMP-2800 Dimatix Fujifilm conductive material printer on Rogers RT/Duroid 5880TM material are displayed in Figure 11. The simulated and measured reflection coefficients, VSWR properties, and their enlarged views are shown in Figures 12 and 13, respectively. The performance of fabricated antennas is measured using an Agilent E5071B vector network analyzer (VNA), and Figures 12 and 13 demonstrate a strong similarity between the simulation and the experimental outcomes.

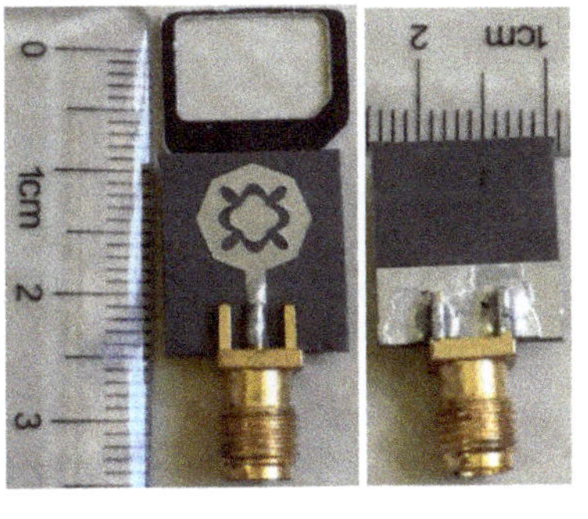

(**a**) (**b**)

Figure 11. Fabricated prototype: (**a**) front view, (**b**) rear view.

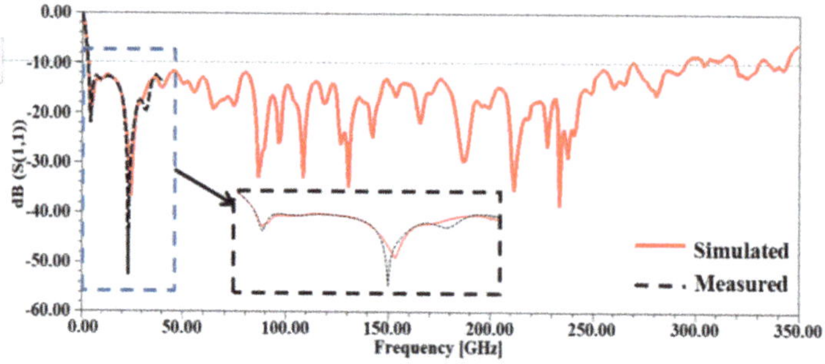

Figure 12. Simulated and measured return loss of the proposed SWB monopole antenna.

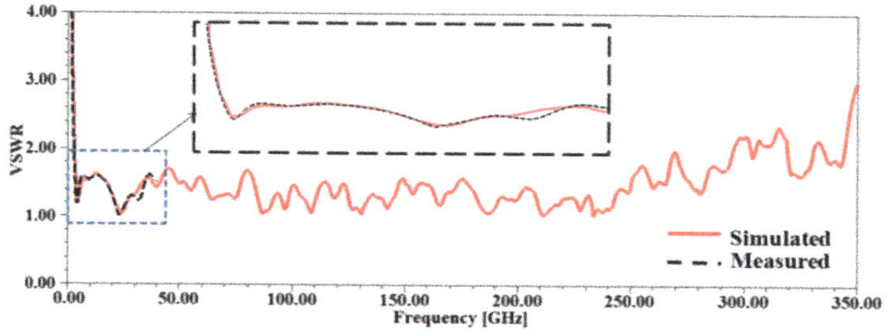

Figure 13. Simulated and measured VSWR of the proposed SWB monopole antenna.

4. Comparison of Proposed Monopole SWB Antenna with Reported Works

The performance comparison of the proposed design and a few recently reported SWB antenna designs [9–24] is shown in Table 4. The proposed antenna exhibits high gain characteristics, excellent SWB response with very high dimension ratio, and a compact size.

Table 4. Comparison to the reported SWB antenna structures.

Ref.	P.S, mm^2	E.S, $\lambda_o{}^2$	f_l, GHz	f_h, GHz	IBW, GHz	FBW, %	BR	BDR	S_{11}, dB	G, dB	η, %
Tiwari et al. [15]	25 × 20	0.20 × 0.16	2.51	16.48	13.97	147.13	6.57	4597.81	−20	5.5	NR
Mondal et al. [17]	58.2 × 47.7	0.97 × 0.8	5	7	2	33.33	1.40	43.23	−22.5	NR	NR
Rahman et al. [16]	57 × 34	0.26 × 0.16	1.42	90	88.58	193.70	63.38	4483.79	−41	7.67	98.9
Singhal et al. [18]	170 × 150	0.37 × 0.33	0.65	35.61	34.96	192.83	54.78	1610.83	NR	6.51	NR
Okan et al. [19]	40 × 40	0.35 × 0.35	2.59	31.14	28.55	169.29	12.02	1419.52	−39	2–5	90–99
Dhasarathan et al. [20]	28 × 20	0.21 × 0.15	2.34	20	17.66	158.10	8.55	5019.11	−20	5.25	NR
Sharma et al. [21]	39 × 39	0.3 × 0.3	2.31	40	37.69	178.16	17.32	1975.62	−40	5.81	NR
Ayyappan et al. [14]	27 × 29.5	0.17 × 0.187	1.91	43.5	41.59	188.51	22.77	5761.87	−62	8.15	90.08
This Work	**14 × 16**	**0.16 × 0.18**	**3.71**	**337.88**	**334.17**	**195.65%**	**91.07**	**6057.27**	**−38.77**	**18.01**	**62–95**

Notation: P.S—physical size, E.S—electrical size, fl—flower resonant frequency, fh—higher resonant frequency, IBW—impedance bandwidth, FBW—fractional bandwidth, BR—bandwidth ratio, BDR—bandwidth dimension ratio, S11—reflection coefficient, G—peak gain, η—radiation efficiency, NR—not reported.

5. Conclusions

A novel portable super-wideband monopole antenna with an octagonal structure was designed for diverse wireless communication applications. The proposed antenna was developed using an octagonal-shaped radiator with microstrip line-feeding. The radiation and bandwidth properties of the antenna were analyzed, and the results demonstrate its suitability for wideband applications. SWB antennas, unlike UWB antennas, are not limited to 3.1–10.6 GHz. Hence, the suggested SWB antenna was compared to reported antenna structures in the literature in terms of physical dimensions, electrical dimensions, impedance bandwidth, bandwidth ratio, bandwidth dimension ratio, fractional bandwidth, VSWR, reflection loss, gain, and efficiency. It is found that a compact SWB antenna is realized and exhibits a very wide bandwidth, a high BDR of 6057.27, and a high realized gain (G) of 18.01 dB with a miniaturized volume. A prototype was developed and measured with the help of the agile vector network analyzer. Even though the proposed antenna operates effectively over the super-wideband frequency spectrum of 3.71–337.88 GHz with a $|S11| \leq -10$ dB, the experimental results are measured only up to 40 GHz due to the higher frequency limitation of the available VNA. The benefits of proposed makes it useful for S, C, X, Ku, K, Ka, V, and W mm wave bands, future-generation mobile networks (i.e., 5G and beyond (B5G), military and civilian applications, cognitive radios for spectrum sensing, radio astronomy for space exploration, satellite communications, and amateur radio for terrestrial and planetary communications.

Author Contributions: Conceptualization: N.S. and S.R.; methodology, N.S. and S.R.; software, N.S.; validation, N.S. and S.R.; formal analysis, N.S. and S.R.; investigation, N.S. and S.R.; resources, N.S.; data curation, N.S. and S.R.; writing—original draft preparation, N.S.; writing—review and editing, S.R.; visualization, N.S. and S.R.; supervision, S.R. All authors have read and agreed to the published version of the manuscript.

Funding: This research received no external funding.

Institutional Review Board Statement: Not applicable.

Informed Consent Statement: Not applicable.

Data Availability Statement: Not applicable.

Conflicts of Interest: The authors declare no conflict of interest.

References

1. Khan, F.; Pi, Z. mmWave mobile broadband (MMB): Unleashing the 3–300GHz spectrum. In Proceedings of the 34th IEEE Sarnoff Symposium, Princeton, NJ, USA, 16 June 2011; pp. 1–6.
2. Rappaport, T.S.; Xing, Y.; MacCartney, G.R.; Molisch, A.F.; Mellios, E.; Zhang, J. Overview of millimeter wave communications for fifth-generation (5G) wireless networks—With a focus on propagation models. *IEEE Trans. Antennas Propag.* **2017**, *65*, 6213–6230. [CrossRef]
3. FCC. *First Report and Order in the Matter of Revision of Part 15 of the Commission's Rules Regarding Ultra-Wideband Transmission Systems*; FCC: Washington, DC, USA, 2002; pp. 98–153. ISSN 1937-8718.
4. Wiesbeck, W.; Adamiuk, G.; Sturm, C. Basic properties and design principles of UWB antennas. *Proc. IEEE* **2009**, *97*, 372–385. [CrossRef]
5. Rumsey, V. Frequency independent antennas. In Proceedings of the 1958 IRE International Convention Record, New York, NY, USA, 20–23 August 1957; pp. 114–118.
6. Balani, W.; Sarvagya, M.; Ali, T.; Pai, M.M.; Anguera, J.; Andujar, A.; Das, S. Design techniques of super-wideband Antenna–Existing and future prospective. *IEEE Access* **2019**, *7*, 141241–141257. [CrossRef]
7. Ali, T.; Subhash, B.K.; Pathan, S.; Biradar, R.C. A compact decagonal-shaped UWB monopole planar antenna with truncated ground plane. *Microw. Opt. Technol. Lett.* **2018**, *60*, 2937–2944. [CrossRef]
8. Chen, K.R.; Row, J.S. A compact monopole antenna for super wideband applications. *IEEE Antennas Wirel. Propag. Lett.* **2011**, *10*, 488–491. [CrossRef]
9. Balani, W.; Sarvagya, M.; Samasgikar, A.; Ali, T.; Kumar, P. Design and analysis of super wideband antenna for microwave applications. *Sensors* **2021**, *21*, 477. [CrossRef]
10. Maity, S.; Tewary, T.; Mukherjee, S.; Roy, A.; Sarkar, P.P.; Bhunia, S. Super wideband high gain hybrid microstrip patch antenna. *AEU-Int. J. Electron. Commun.* **2022**, *153*, 154264. [CrossRef]
11. Samsuzzaman, M.; Islam, M.T. A semicircular shaped super wideband patch antenna with high bandwidth dimension ratio. *Microw. Opt. Technol. Lett.* **2015**, *57*, 445–452. [CrossRef]
12. Dey, S.; Arefin, M.S.; Karmakar, N.C. Design and experimental analysis of a novel compact and flexible super wide band antenna for 5G. *IEEE Access* **2021**, *9*, 46698–46708. [CrossRef]
13. Barbarino, S.; Consoli, F. Study on super-wideband planar asymmetrical dipole antennas of circular shape. *IEEE Trans. Antennas Propag.* **2010**, *58*, 4074–4078. [CrossRef]
14. Ayyappan, M.; Patel, P. On design of a triple elliptical super wideband antenna for 5G applications. *IEEE Access* **2022**, *10*, 76031–76043. [CrossRef]
15. Tiwari, R.N.; Singh, P.; Kanaujia, B.K. Small-size scarecrow-shaped CPW and microstrip-line-fed UWB antennas. *J. Comput. Electron.* **2018**, *17*, 1047–1055. [CrossRef]
16. Rahman, S.U.; Cao, Q.; Ullah, H.; Khalil, H. Compact design of trapezoid shape monopole antenna for SWB application. *Microw. Opt. Technol. Lett.* **2019**, *61*, 1931–1937. [CrossRef]

17. Mondal, T.; Maity, S.; Ghatak, R.; Bhadra Chaudhuri, S.R. Design and analysis of a wideband circularly polarised perturbed psi-shaped antenna. *IET Microw. Antennas Propag.* **2018**, *12*, 1582–1586. [CrossRef]
18. Singhal, S.; Singh, A.K. Elliptical monopole based super wideband fractal antenna. *Microw. Opt. Technol. Lett.* **2020**, *62*, 1324–1328. [CrossRef]
19. Okan, T. A compact octagonal-ring monopole antenna for super wideband applications. *Microw. Opt. Technol. Lett.* **2020**, *62*, 1237–1244. [CrossRef]
20. Dhasarathan, V.; Sharma, M.; Kapil, M.; Vashist, P.C.; Patel, S.K.; Nguyen, T.K. Integrated bluetooth/LTE2600 superwideband monopole antenna with triple notched (WiMAX/WLAN/DSS) band characteristics for UWB/X/Ku band wireless network applications. *Wirel. Netw.* **2020**, *26*, 2845–2855. [CrossRef]
21. Sharma, V.; Deegwal, J.K.; Mathur, D. Super-wideband compact offset elliptical ring patch antenna for 5G applications. *Wirel. Pers. Commun.* **2022**, *122*, 1655–1670. [CrossRef]
22. Kundu, S.; Chatterjee, A. A compact super wideband antenna with stable and improved radiation using super wideband frequency selective surface. *AEU-Int. J. Electron. Commun.* **2022**, *150*, 154200. [CrossRef]
23. Singhal, S.; Singh, A.K. CPW-fed hexagonal Sierpinski super wideband fractal antenna. *IET Microw. Antennas Propag.* **2016**, *10*, 1701–1707. [CrossRef]
24. Waladi, V.; Mohammadi, N.; Zehforoosh, Y.; Habashi, A.; Nourinia, J. A novel modified star-triangular fractal (MSTF) monopole antenna for super-wideband applications. *IEEE Antennas Wirel. Propag. Lett.* **2013**, *12*, 651–654. [CrossRef]
25. Mythili, P.; Das, A. Simple approach to determine resonant frequencies of microstrip antennas. *IEE Proc.-Microw. Antennas Propag.* **1998**, *145*, 159–162. [CrossRef]
26. Ray, K.P.; Kumar, G. Determination of the resonant frequency of microstrip antennas. *Microw. Opt. Technol. Lett.* **1999**, *23*, 114–117. [CrossRef]

MDPI
St. Alban-Anlage 66
4052 Basel
Switzerland
www.mdpi.com

Applied Sciences Editorial Office
E-mail: applsci@mdpi.com
www.mdpi.com/journal/applsci

www.ingramcontent.com/pod-product-compliance
Lightning Source LLC
Chambersburg PA
CBHW062033220926
51850CB00010B/2536